Modernity & Technology

Modernity & Technology

Harnessing the Earth to the Slavery of Man

WADE SIKORSKI

The University of Alabama Press
Tuscaloosa & London

Tuscaloosa, Alabama 35487–0380

Manufactured in the United States of America
designed by zig zeigler

The paper on which this book is printed meets
the minimum requirements of American
National Standard for Information Science-Permanence
of Paper for Printed Library Materials,
ANSI Z39.48-1984.

Library of Congress Cataloging-in-Publication Data

Sikorski, Wade, 1956–
Modernity and technology : harnessing the earth to the slavery of man / Wade Sikorski.
p. cm.
Includes bibliographical references and index.
ISBN 0-8173-0667-6 (alk. paper)
1. Technology—Philosophy. 2. Ecology. I. Title.
T14.S54 1993
303.48′3—dc20 92-16837

British Library Cataloguing-in-Publication Data
available

For my mother,

LUCILLE

Contents

Acknowledgments

N MANY WAYS this book is indebted to the help of other people. My mother, to whom this book is dedicated, was always there in kindness for me in ways too numerous to list. My father, Edward Sikorski, and my uncles Jerry and Tim Sikorski taught me much that helped me think the thoughts in this book.

My grandmother, Rose Griffith, generously gave me money and household appliances. My sister, Sandy Sikorski, tirelessly ran off multiple copies of anything I needed and did many small favors for me that only she could do. Many of my friends from the University of Massachusetts, especially Jane Bennett, gave me encouragement and sent books and papers, and hope, when I didn't have much. Bill Chaloupka generously advised me on portions of the manuscript.

While I was at New Mexico State University, Darian Goldsmith and Diana Kovar worked hard and patiently to help me prepare this manuscript. Bonnie Bright helped me work with publishers. Russ Winn generously helped me set up my computer. Discussions with Victor Aikhionbare, Nancy Baker, Ann Beck, Bonnie Bright, Sam Combs, Gloria Friedman, Leonard Gambrell, Ed Hall, Y. K. Hui, Lisa Johnson, Sadie Leach, Alynna Lyon, Pat Michaels, Valerie Miller, Helen Quintana, Lois Richards, JoAn Rittenhouse, Jim Seewald, Dimitri Stevis, Mary Wolf, as well as many students of mine, too numerous to mention, helped me rethink the ideas in this book.

As the author-ities whose reading this writing first awaited, anticipated, and dreaded, the skills of Bill Connolly, who was my dissertation chair, Jerry King, and Mike Best are re-presented in this work. I hope that it is worthy of their efforts as teachers and thinkers.

And finally, my thanks to Malcolm MacDonald, Craig Noll, and the rest of the staff at The University of Alabama Press. Thanks also to all the anonymous readers. This book has greatly benefited from their editorial guidance, their craftsmanship, and their careful and thoughtful reading.

Modernity & Technology

CHAPTER ONE

Thinking *about* My Place

my reader. . . .

Lost in dreams of you
Thinking of you
Reading me
your sky your goddess
builds me
your thinking your earth

hoping your hopes
fearing your fears
caring your cares

for what am I
but your reader
and your dream?

—the author

IN THE PASTURE behind my home there are still traces of how people used to live on my place. Just a ways down a creek full of cattails and reeds, an old farmhouse, faded gray and falling down, stands near a small pond and a dying tree. Beside it, barely visible through the growing grass, are the foundations of other buildings—a granary, a blacksmith's shed, a woodshed, and perhaps a barn. Farther away, a line of rhubarb plants still struggles against the prairie grass, probably near what used to be a garden. Farther away still there is a shelter belt of aged and slowly dying cottonwood trees, maybe sixty feet tall.

A family used to live here, but now the cattle have pushed into their old home, seeking shelter from the winter storms. They have stomped the floorboards into the ground, rubbed against the supporting braces, knocking down walls and leaving strands of their hair on the nails that stick out. The brick chimney has collapsed, leaving a hole in the roof for the rain, the snow, and the wind to come in. Soon, the entire building will fall to the ground, leaving the cattle without a shelter.

The soil around this old farm is sandy. In the thirties, when the drought and the grasshoppers came, it blew. Badly. Where there was once wild and lush prairie, a home to buffalo, prairie dogs, coyotes, and Indians, shifting sand dunes grew, rolling and crashing like a storm-tossed sea behind the plow. Now, the grass grows only in clumps almost a foot apart, so fragile that one can reach down and pull it up by the roots with one easy jerk. The thick rich sod of the prairie has been replaced by scattered desert plants, cactus and yucca. Only a few years ago have some of the worst blowouts grassed over enough to stop the blowing. Now, depleted, exhausted, this old farm is a winter pasture for our cattle; the people who lived here have left, probably for the city.

There are many old farms like this on my family's ranch in southeastern Montana. We remember them by the names we call places—the Chapman place, the Morton place, the Pepper place, the Blazer place, the Sawyer place, the Harris place, the Jones place, the Hough place, the Frankie place, and the home place. And perhaps there are a few places whose names we have forgotten. Those all were farms and homes that my family took over when the land would no longer support them. When I was a little boy, we had one of the largest ranches in Fallon County. Now, though we have sold none of our land and have even bought some more, most of our neighbors are bigger than we are.

Perhaps one day, following this "natural" progression, it all will simply become the Sikorski place, and the names of all the places my family remembers will be forgotten, like the names of all the places the Indians remembered.[1]

The Reagan administration, and now the Bush administration, following the truth of our time, calls this progress. The inefficient and nonproductive are swallowed up and dis-placed by the more efficient and more productive, and the whole economy is made more rational as a result. Resources—human as well as nature's—are recentered, redistributed, and used in a way that maximizes their utility for a global economy. Large scale is more efficient and more productive, more capable of rendering up nature as a resource for the economy, and so it is more rational. Who but a poet can be so sentimental to doubt this truth?

Disciplined by the harsh realities of the free market, American farmers are very productive, and becoming more so every year. According to the USDA, one farm worker now supplies enough food and fiber for seventy-six people, whereas ten years ago she was only producing enough for fifty.[2] One hour, we are told breathlessly, of farm labor today produces sixteen times as much food as it did in 1920.[3] Where once, and not so long ago, the vast majority of Americans were engaged in farming, now only 2.5 percent are.[4] The rapid expansion of farm productivity freed people for other things—working in the factories that made tractors for the farmer, the transportation system that hauled the farmer's product to market and distributed it, the financial system that loaned the farmer the money to buy the new technology, and the chemical industry that supplied the farmer with the means of controlling pests.

Fertilizer use grew fifteen times from its 1930 level to now, tractor horsepower eleven times, the number of tractors five times, and chemical use from nothing to its present levels of saturation.[5] As a result of all of this, according to the USDA's fact book, farming now uses more petroleum products than any other single industry in America—and that does not include the energy necessary for the distribution, preservation, and consumption of farm products.[6]

To get produce from the farm to the consumer's table, hundreds of billions of dollars are spent each year assembling the food, inspecting it, grading it, storing it, processing it, packaging it, wholesaling it, and retailing it. It travels across 177,400 miles of railroads, 3.2 million miles of intercity highways, and 26,000 miles of improved waterways.[7] Besides the farmer, the American food system directly employs 20 million people to transport, process, and sell the farmer's products—approximately one out of every five jobs in private enterprise.[8] This, however, still reflects only a part of America's food system. It does not include the people whose work indirectly supports it. It does not include, to mention but a few, the people employed supplying the entire food system with energy (oil, electricity, coal, and nuclear power), nor does it include the people employed building and servicing the transportation system (the railroads, trucks, waterways, ships, tires, and highways), nor does it include the people employed building and maintaining the refrigeration systems necessary for stores, shipping, and home use, nor does it include insurance and financial costs for these subsystems, and it certainly does not include all of the external costs of all these operations—the pollution of the land, air, and water, the extinction of various species, the destruction of buildings and monuments, poor health stemming from pollution, chemical additives to food, and so on.

That modern farming is more productive than earlier ways, a great triumph for humankind, is a truth that is possible only for analytic reason. Describing farming as only the simple production of food, measuring productivity as only the relationship between input and output, and knowing progress as the continual expansion of human control over this relationship, analytic reason assumes that farming is nothing more than simply growing food for as many people as efficiently as possible. Traditionally, the farm household not only grew food but processed it, stored it, and transported it to where it was used. Drawing near to itself everything that the modern economy spreads over entire continents, the traditional farm household was almost entirely self-sufficient.

Some idea of the true extent and interdependence of America's food system is contained in the fact that a typical household, now in no way a farm household, spends close to a third of its income directly on food.[9] If we can generalize from this cost, then, we could say that America spends almost a third of its working effort supporting the food system. This, however, is still an inadequate measure of its interdependence. Not only is it dependent on a global food system, the modern household must be linked up to a variety of systems to participate in the general food economy, among them the water system, the sewage system, the transportation system, and the electrical power system. It is necessary to own a car in order to get food from the distant shopping mall; it is necessary to be hooked up to the electric grid to preserve the food brought home with refrigeration and to cook it; and it is necessary to be hooked up to the municipal water and sewage systems to get rid of the resulting waste.[10]

Since the modern household must be hooked up to these systems, the costs of participating in the modern food system go far beyond what the modern household directly spends on food. If we include the unpaid work of the housekeeper with all the direct and indirect costs of participating in the modern food system, we can only conclude that the modern household devotes a very considerable effort to feeding itself—way over half its time.[11]

Perhaps it is time we asked if the modern food system, as a whole, is really that efficient and productive, if the reason that has built the modern food system is that thoughtful? It is one thing to employ sophisticated and complex technology; it is an entirely different thing to bring it close to where we live. Displaced, vast, and unconnected, the entire food system seems to have been built without any purpose at all in mind. Of course it feeds us, but does it do it well? Does it enhance our lives by cultivating a sense of beauty, awe, mystery, community, and health? Does it connect us with each other, the earth, the sacred, our own life possibilities? Or does it just feed us, leaving us alone, afraid, and apathetic, indifferent to each other and the earth?

A rhetorical question. But think, dear reader, of what would happen if instead of using this vast food system, if instead of all of this huffing, puffing, and pollution, we grew our food in a garden near our house, carried it by hand to a root cellar for storage, then carried it to our kitchen when we needed it, and recycled the wastes to our garden?[12] We could not possibly spend more of our time feeding ourselves than we do now, and we would not have to deal with all the results of the modern food system—the pollution, the health risks, the security risks, and the degrading work of the factory.

Perhaps this way of living could give us quite a bit more than just food. Gathering near at hand what modern technology has dis-placed and made distant, we could have a situated way of living to replace the dis-placed and purposeless rationality of our economy with the governing care of friendship, community, health, and wholeness. Instead of being imprisoned by the pointless and distant logic of the marketplace, the industrial economy, and government regulation, we might respond with care to the earth, the changing seasons, and the needs of those near to us. Maybe.

The dis-placement of the family farm, a largely accomplished fact, embodies the dis-placement of our thinking, for drawing near to the cares at hand in our lives.[13] Forgetting its place and so its purpose, we allow the necessity and truth of technical efficiency to draw us up into vast systems of specialization, rationality, and control, imprisoning us within the logic and rationality of an economy that dispossesses us of our land and our homes. Then, we become lost to the cares of life, and modern technology makes us into rational, unthinking robots indifferent to each other, the earth, and the sky above. It destroys our homes, forgets our memories, and makes the whole world desolate with its indifference.

Am I too angry and accusing? Too soon to judgment? Perhaps. But wait, my reader, there is a place for anger . . .

The reasoning that denies us our anger does so out of an imperative to be universal, out of a requirement to level out every local or temporal difference and possible moment of incommensurability and to assert unbroken control over it, fixing it within a timeless, impersonal, and placeless organization of objective truth.[14] Something that is never angry or lost to love —that is what truth is for modern reason, something that is not situated by bodies or cultures, times or places. It must be beyond all of that. Posited by Man, the discursive formation that governs this age, for his utility, modern reason projects a crystalline pure conceptual grid over all places, locking them tightly in its logic, re-presenting them as Man's own utility. Because it derives itself from no place, just Man abstractly imposing his will on the world, mathematics and especially geometry are the purest expressions of modern reason. In geometry, formal and exactingly precise definitions or representations are first posited (a point, a line, the meaning of parallel lines), and then these simple and entirely unambiguous definitions and representations are used to fabricate proofs, building up a complex grid of truth through which the world can be interpreted.[15] Through these representations, reason seizes hold of the world and makes it into Man's utility.

Unlike reason, which derives its logic from a universal human utility, thinking is a handcraft—a handcraft because the truth it knows is situated, a temporal and a local response to the things that are near at hand. Coming forth in poetry, dance, art, and song, thinking is not any mere human instrument, governed by the utility of Man's dominion over things; it does not seek control over things, nor does it require universality. It erupts as the joy of life, a calling seeking meaning and beauty. Opening itself up to the eruption of truth amid the mystery of life, thinking reveals only itself. Unconcerned with principles, laws of logic, or universal applications, thinking is a gentle response, a meditative awareness that reveals the mystery of being situated, of having a place. It is the poetry and song of here and now, this language, this body, this culture. It is an openness to this place, the world that worlds here, the thing that things here, not an aggressive universal and eternal positing that indifferently imposes itself on all places.

Building its world with what is near to it, it is a handcraft because it responds to the particular thing at hand as a temporal and local event, letting it guide it on its way toward an unconcealment of Being. As thinking goes its way, step by step drawing nearer to the temporality of the thing rising up out of nothingness, it reveals its mystery. The mystery that a thing things here, and the mystery that it could also, because presence always brings absence, not thing here. Gently letting Being be, thinking does not build large systems, freezing words in a formal and eternal structure that tolerates no imprecision, ambiguity, or play. Governed by the poetry of its place, responding to the song that is close to it, thinking forever starts anew as the thing at hand changes, using words anew as their context changes.

As all poets, artists, and dancers know, there are no eternal rules that guide the craft of thinking to its truth, but rather there are the ways that the great masters of the craft—each according to their time and place—have cleared. As each new generation of thinkers follows these paths, they learn they must eventually make their own. Drawing near to their own life, their own place and time, the craft knowledge they have gained will call on them to depart from the ways of the master craftsman because the thing at hand, rising up before them in all its mystery, calls them to its own time. Because thinking comes from dwelling, and because dwelling places, being local, temporal, and near at hand, are never the same, the craft of thinking is never practiced the same way when it is done as it should be done, not even in our time, when the same presses upon us with such monotonous regularity.

The great danger of our age is the flight of Man, the master of reason, from thinking, from responding with poetry and song to the eruption of the thing; his great doom is his failure to recognize his loss.[16] The failures of reason are as dangerous and blinding as its triumphs are powerful and revealing. Never before has any culture been as ambitious or had so many plans as ours, launched so many scientific inquiries into the design of nature, or attained such great mastery over it. No one can deny that the physical sciences that made possible our nuclear power plants, our computers, our space shuttles, and our communications satellites are great, unprecedented, and of awesome power. And it seems likely that our life sciences, now that they are unraveling the possibilities of recombinant technology, promise even greater triumphs. Nor is it necessary to repudiate all this knowledge and power and, just throwing it away, to return to an earlier, simpler time.

More important is the calling of poetry, the calling to think about what all this knowledge and power is useful *for*—what use it is to the place we live at, what truths it serves. Friendship? Community? Art? Beauty? The gods? Or is power just serving itself—aimless, pointless, wandering, and lost? As awesome as all of this power is, all of it is but calculation and counting, mere reason that does not meditate deeply on the meaning of everything that is or does not draw it near to the dwelling place. Indifferent to distance, pain, isolation, or the love it sacrifices, reason computes, and it computes even if it does not use numbers, a calculator, or a mainframe computer. Never stopping, never collecting itself, never asking why, reason jumps from one solution to the next, forever seeking more efficient or economical equations, just because they are more "efficient" or "economical." It is the exclusive mark of our age, the source of its progress and its mastery over everything.

But if mere reason is justified and needed in its own way, if it is useful and handy, it has not found its place or its time. Not permitting meditation or thinking to guide it on its way, it recklessly and carelessly spreads its force over all the earth, callously reducing anything that had escaped it to the black-on-white monotony of its equations. For it, meditation and thinking are but colorful voyages in fantasy, utopian daydreaming, floating off in a realm useful only for poets and perhaps the entertainment industry.[17] For it, poetry, song, and dance may be pretty and may describe a wonderful world, but they are useless in this vale of tears—worthless for the current business of production and out of touch with practical affairs. Furthermore, demanding such excellence, it elevates itself so far beyond common understanding to be available to none but a tiny and effeminate elite of poets, mystics, hermits, artists, and nuts that do nothing but meditate all day long. It is utopian, in other words—hopelessly so.

In fact, reason, by raising up placeless truth, objective and universal, is far more utopian than thinking. Utopia was the name Sir Thomas More gave in 1516 to a fantasy island, supposedly perfect in all moral, social, and political matters. In New Latin, *utopia* means "no place," a place that does not exist, a placeless place, or, more generally, an abstract fabrication made for the purpose of edification.[18] Such is in fact the true character of reason, building, as it does, fantasy islands made of numbers and logic in order to control the world —not of thinking, however, which is always firmly rooted in the earth, in the feelings of the body, in the soul-stirring callings of love, enchantment, mystery, and beauty. When reason denounces thinking for being utopian, it is only projecting its shadow onto its other, thereby denying itself its own reality.

This dismissal that calculative reason makes of meditation, poetry, art, and thinking reveals more about itself than it does of the alternatives it dismisses. It does not acknowledge that meditative thinking is a handcraft, an art that responds only to what is near to its time and place. It is not irrelevant, a luxury, or a hobby but the most relevant practice of all—to any time, to anyone. It draws life near, makes it meaningful and beautiful, and, as such, is as available, or unavailable, to common understanding as it is to highly practiced understanding. So long as they live, mortals cannot escape the calling that thinks in poetry, art, dance, and song. While they do require great effort, much practice, delicate care, and occasionally great courage, they always call upon common people as much as any elite. Though it can be ignored, the calling to poetry, dance, art, and song is simply life.[19]

Unlike calculative reason, which requires specialized technical education, meditative thinking is done—and must be done—by anyone living their life. This is because, as dwellers on the earth, responding to that which brings them forth and keeps them safe, human beings are thinking beings, even if in our time they so often neglect to think. Thinking is the way we dwell, the way we let the world world. To think, it is always enough to dwell on what lies near and to meditate on what lies nearest, enough to think about whatever concerns us in our neighborhood, in our time.[20]

But thinking is discouraged in this age, repressed, denied, and even persecuted.[21] Reason has built our world, and it has done it in a totally thoughtless way, a way heedless of poetry, dance, art, and song. It has built our economies, our weapons, our houses; it governs the way we grow our food, raise our children, even create our art—everything. And now everything is trapped in the utopia that reason has built. Projecting itself over everything, drawing everything into the domain of its placeless truth, reason has fabricated its own fantasy and made us live it. Fearing any challenge to its perfection, it must extend itself continually, mastering everything that escapes its necessities—irrationality, unplanned behavior, local dissent of any kind.

Because of their placeless rationality, their unending quest for impersonal standards, objective judgment, and logical procedure, the modern bureaucracy, the modern state, and the modern economy cannot function if people think about what they are doing or respond to the cares of their lives, such as friendship or beauty. As parts of an industrial machine with a global reach, people must do their narrow function within the system, orders must be followed unquestioned, and the rationality of the huge and far-flung system must prevail over any local difference, any calling from poetry or song.

The modern age has thus given rise to a new experience of evil that is so terrifying because it is so possible, so near to us all. Adolf Eichmann, like all us conscientious workers, was just doing his job, fulfilling his function within the system—only his was herding Jews into ovens.[22] An efficient technocrat, he might have been disposing of day-old newspapers, baby diapers, or toxic wastes, but he was disposing of Jews. It made no difference to him; there was no personal hatred in his intentions. Why, he even had Jewish friends!

If Eichmann were truly an exception, a shadow that was not our own reality, his life could not be a warning to us. But perhaps Eichmann was not an exception. Perhaps he is our collective shadow, our own truth repressed into the unconscious and then projected safely onto another. All over this country, and all over the world, there are missile silos armed with weapons so deadly that they can do in a moment what it took Eichmann years to do. They are manned by men and women carefully selected, as the logic of deterrence theory dictates, to obey their orders and push a button that will kill millions of people the moment they receive a code word.[23] No doubt these men and women are better than most of us. They are more intelligent and better disciplined; they have no vices, love their children, and never cheat on their taxes. But yet, with just a few code words, they may one day become the instruments of global destruction. And we participate with them in this possibility with at least our silence, our taxes, and sometimes our votes.

In this age that reason has built, it is not easy or safe to think about what we are doing, to respond to the cares near to our lives. Fearing the identity given to those others that reason excludes, it is much easier to be reasonable and go with the flow, to work inside the system, to do one's duty. The penalties for the courage to think our thoughts to their end, to integrate reason's shadow and draw our thoughts near to the life that sustains them, are so great—ignominy; loss of career, home, and family; perhaps even jail and death—that few of us have the strength necessary for thinking them. The shadow rules.

Fearing the passionate dangers of poetry, many flee into such things as fundamentalist religion, positivism, and rational-choice theory. Reducing everything to fundamentals, a simple distribution of the world into black and white, quantity and quality, logic and irrationality, method and error, right and wrong, such attitudes eliminate the greatest of all modern demons—the uncertainty of whether God, objectivity, and reason are real—and leave people secure in their utopian islands. Everything is OK. In a time when people have been cut off from their roots, their homeland, the earth that sustains them, and they wander about the world in trailer-houses, they are haunted by the most horrible of fears. Unconsciously they know everything is not OK. Isolated by their utility, dependent on the world that reason has built, afraid of what would be released by any real questioning, they seek security from their unconscious doubts in absolute and eternal certainty. The utopia of reason is a place haunted by a horrible shadow.

Our fear, as we live amid the institutions that reason has built, is a fear of reason's shadow, uncontrollable chaos, ambiguous uncertainty, monstrous unreason, erupting contingency. Pursued by its shadow, everything that it is not but nevertheless is, reason calls us to ever-expanding conquests in the cause of reason.[24] For the security of all that we have built and then become dependent on, we must repress the shadow of our age and make everything rational, available for human control. Otherwise mere anarchy is set loose.

In Jungian psychology, the encounter with the shadow is often experienced as a confrontation with a dangerous beast, some horrible monster that seeks the self's destruction. The self feels like a helpless victim before the onslaught of the unconscious, even though this monster is made possible and constituted only by its own ego ideal. Often the self protects itself from the judgment and hostility of the ideal by projecting its faults onto others, rejecting its own power. This other assumes the ego ideal's own aggressive, predatory, and evil character, and then the other sets about persecuting the self. Healing comes when the self withdraws its projections, recognizing that all are one, hunter and hunted, judge and accused, executioner and condemned, jailor and prisoner, that it is its own ideal that is persecuting it, not the other's.[25]

Insisting on the purity of its analytic categories, reason is singularly unable to face its own demons and recognize them as its own. Its faults are always projected onto others, whatever is different from itself, whether Communists, capitalists, terrorists, feminists, environmental activists, drug dealers, or whatever. The rules and methods of reason are fixed, resolute, unchangeable. Anyone who ceases to follow them ceases to be reasonable and thus becomes the enemy. Consequently, while thinking, paying attention to the erupting earth, actively seeks out alterity, its others, and knows its identity only in the play of difference between world and earth, reason is unable to overcome its projections of difference and heal itself. Whenever it is threatened, fearing for its identity, it simply renews its efforts to protect itself from its projections. Ultimately, everything that reason knows is known only because of its fear, hatred, and insecurity. If reason insists on its dispassionate objectivity, its calm logic, and its peaceable ways, it is able to do so only because it has given so much of itself over to its enemy. Concealing and repressing its fear and hostility, lost in fantasies of persecution, frantically needing always more power to protect itself, reason is, despite its endless protests to the contrary, a profoundly warlike and hostile way of building worlds.

It is no accident that reason has built the nuclear bomb and has described it as a way to keep the peace. "Deterrence theory" it is called. But even with it and tens of thousands of nuclear bombs, we are not safe. Even now, when we in America have had the comfort of seeing our enemy, the Soviet Union, fall into disarray, some still counsel continuing vigilance, caution, and deterrence. And above all, they would allow only small cuts in the defense budget, even when our schools are an international disgrace, when black infant mortality rates rival that of Third World countries, when the homeless crowd our streets, when the ozone layer is being depleted faster than anyone expected, when the greenhouse effect is building up, when toxic waste dumps are seeping into the nation's water supply, when species after species are becoming extinct, when farmland is being eroded much faster than it is being rebuilt, and so on. Even with all these problems, still we have enemies to fear, to deter, to spend huge amounts of money on. And no one is even sure who they are anymore.

Even when everything has changed, nothing has changed because our enemy never was the Soviet Union. Not really. The unrelenting hatred for democracy, life, and freedom and the unqualified quest for world domination that would cause our enemy, the Soviet Union, to launch nuclear war against us were mostly our projections. Who could be that monstrous?

Not a few nations, especially in the Third World, have seen in American foreign policy everything it has denounced in the Soviet Union. When they think the unthinkable, the actual strategy and execution of a nuclear war, our generals should be reprimanded not for the insane courage of thinking thoughts too horrible to bear but for the cowardice of not thinking at all. Deterrence theory is a masterpiece of rationality, calculation, and logic, but it is also a failure of thought and courage.[26] And it is so distant from our daily cares and concerns that it hovers over us like an alien spacecraft. It posits an artificial and utopian world of "exchange rates," "intense interactions," "counter forces," and "counter values," and it requires that the fear that its fabrications conceal govern the real world.[27]

The enemy our generals seek to protect us from lives only in the Pentagon. But this dwelling place of the enemy does not make it unreal. On the contrary, the enemy that must be subdued is all too real. The truth of his presence is attested to by the tens of thousands of nuclear warheads and delivery systems we have built and, more ominously, by the war bureaucracy whose decision procedures would use them on a moment's notice.

Our generals assure us, in moments when our resolve to face the nuclear age weakens and crumbles, that they have made war impossible because they have made it too horrible. The enemy is contained, subdued, and made rational by the terror they, the masters of an insane terror, inspire. But this promise of safety ignores the fact that large numbers of nuclear weapons still exist, also, that the decision procedures and control centers for using them exist—and will continue to exist as long as reason feels threatened by its others. Only such power as this is capable of protecting reason from its fear, and it will endure as long as reason is lost to its enemy.

There is perhaps another way to do things, to know things, a way that is not lost in a world far removed from itself. Believing there was a way other than the way of reason, I tried to find it by focusing on my place. Just my place. Up the hill from the old abandoned farm, set deep in a hill that overlooks a valley a dozen miles wide, I built an earth-sheltered house and greenhouse. Not needing much energy (none for heating, even in Montana's worst winters) and growing my own food, I can live here for a year quite comfortably on only a couple of thousand dollars. I am therefore blessedly free from the demands and sacrifices of continuous employment. (For years now, I have not made enough money to pay any income taxes. I console myself in my poverty by thinking of how little I have contributed to the nuclear arms race.) Though I get lonely sometimes, and though there are sometimes plagues of grasshoppers that sweep down off the prairie and eat my garden to the ground like a horde of lawyers on a medical malpractice suit, I can live my life here near to what sustains it. It is a life sometimes full of petty frustrations, I must admit, and it is not always a happy life, but it is my own. And I have learned a lot here on my place, things that I could never have learned elsewhere.

Sometimes late at night I wake from my sleep, afraid for some reason I cannot really describe. I put on my clothes and wander about the hills that surround my house. One of them is the highest for many miles around. On moonlit nights when I stand on top of it, I can just see in the distance some hills that conceal a number of Minuteman missile silos. I remember driving past them. They are not much, really, just a woven wire fence around some concrete slabs. But their presence is never far from my mind on these nights. Sometimes I take my binoculars along and I try to make them out in the moonlight. Is it that hill that hides one, or is it the next? I am never sure.

I give up and look up at the night stars, see them sharp and cold against the blackness. And I think again how wonderfully indifferent they are to our hopes and fears, and how much they are like death, cold and certain. Strangely, thinking of death like this on a starry night, alone and shivering on a hilltop, is comforting. Perhaps because it seems so small against the stars. I think of the Indians who have stood where

I stand, looking at the same stars. Surely they must have done so. And I think about how they are gone now, with their whole world of buffalo, medicine, and grass. And I know then that nothing lasts as long as the stars—not the buffalo, not the Indians, not my family's ranch, not me, not the Minuteman missiles, not even the stars last forever. I go home and get back into bed, no longer afraid.

CHAPTER TWO

Thinking *about* Technology

> The white man, through his insensitivity to the way of Nature, has desecrated the face of Mother Earth. The white man's advanced technological capacity has occurred as a result of his lack of regard for the spiritual path and for the way of all living things. The white man's desire for material possessions and power has blinded him to the pain he has caused Mother Earth by his quest for what he calls natural resources. And the path of the Great Spirit has become difficult to see by almost all men, even by many Indians who have chosen instead to follow the path of the white man.
>
> —letter from Hopi Indian leaders to President Nixon

IN HIS THINKING about technology, the Thinker seeks to prepare the way to it and make it possible for humanity to have a free relationship to it and to itself.[1] The first step along this way is to distinguish, to think of the nonidentity, between technical activity and the true character of technology. They are not the same. Just as the identity of a tree, as tree, is something that pervades, precedes, and makes possible all trees but is not itself a tree, the identity of technology has nothing at all to do with technical activity.[2] That which makes technology possible, which brings it forth and gives it its distinctive identity, cannot be known by pushing forward the domain of mechanics and calculation or by attempting to conclude, as the popular debate over it does endlessly, that technology is either good or bad, to be extended or contracted, or, much worse, that it is something neutral, merely a tool depending on humanity's virtues or failings for its good or bad effects. Such a way of proceeding would not enable us to think its truth or reveal what brings it forth but only deliver us over to its rationality more completely.

To seek the truth of a thing, that by which the thing things, is to ask what it is, to think about the way in which it is present and absent. In our time, two answers are commonly made by rational discourse on what technology is. The first says that technology is but a means to an end, a way of asserting power over things; the second says that technology is a human activity, a way of asserting human control over things. According to the Thinker, these two common definitions supplied by reason belong together, for to name ends and seek the means adequate to them is a human activity, exclusively so in our time.[3] Not only are machines, tools, and factories part of what modern technology is according to modern reason, but so too is the end, the human will they serve.

Modern humanity gains its power, its mastery over things, and accomplishes its will through its technology. Thus, the Thinker calls this rationalist conception of technology the *instrumental* and *anthropological* definition of technology.[4] There is nothing incorrect about this identification of technology as a bringing about and as human power as far as mere correctness goes. In fact nothing is more true of it than in what is absent from it, concealed within it. It is an understanding entirely appropriate to this age of reason.

This concept of technology as a bringing about and as human power defines it as something universal and, as such, able to interpret not only ancient and premodern technology but also modern technology, including not only the tilling sticks, the water and wind mills, and the horse-drawn implements that the ancients used but also the nuclear power plants, the four-wheel-drive tractors, the jet aircraft, and the big factories that we use. All of these techniques, reason knows, are means to ends, ways of bringing things about and asserting human control over them, even if they differ greatly in power and complexity.

But again as reason knows, if modernity has only expanded the power of technology and extended Man's mastery over the earth, it has also thereby greatly increased the need to get technology itself under control, to master it just as Man masters the earth through it. The great fear of reason is that the means by which it imposes its will upon the earth and all that is in it will escape human control, as Frankenstein's monster did, and subject the new Prometheans to a cruel fate—perhaps nuclear war, perhaps ecological disaster. The will to technical mastery becomes more urgent and insecure just as it becomes more powerful and sure of its domain. Reason's shadow rules it unconsciously, making it endlessly more fearful of the world it might lose control of, provoking it into ever-expanding conquests.

The modern will to mastery must find in itself only itself: pure will willing itself. No shadows allowed. Anything that is not reason itself, escaping its control, must be isolated, contained by knowledge, and made into a yet greater means of control. Technical Man must find in his will the cause to all the effects his technology brings about. Anything else is an enemy to be opposed, overcome, subordinated, mastered. Wherever reason and the instrumentality of technology reign, there reign causality, power, domination, and human subjectivity.

But, the Thinker inquires, what if in its truth technology were no mere means, not simply a human way of bringing things about, universally neutral as long as Man had control of them, simply more power to make the world submit to Man's choice?[5] What if Man never did have control over the technology he deploys and could not bring things about by willing them but rather was in his willing always preceded, appropriated, and possessed by something other than his will, something before it, more primal than it? What if Man's will could not be a cause, and technology a mere means to human ends, and Man could not choose, but was in fact chosen? What if the shadow that Man had long opposed, repressed, and sought to master was in fact not defeated but his master? The Thinker responds by probing the nature of causality and human willing, finding that it is quite other than reason has long maintained.

Before the age of reason, philosophy had long maintained that there are four different kinds of causes to anything that occurs and becomes present:

1. the *material* cause, the physical substance out of which a thing is made;
2. the *formal* cause, the form or shape into which the material enters;
3. the *final* cause, the end for which the matter is pressed into its form;
4. the *efficient* cause, the actual force that brings about the effect.[6]

For example, the material cause of a silver chalice would be the silver that went into it; the formal cause would be the cuplike form it became—smooth, gilded, or whatever; the final cause would be the sacrificial rite to the gods that it was made to participate in; and the efficient cause, in this case, would be the craftsman who made it.

In the age of reason we have become accustomed to representing cause as that which brings something about. Efficient causality, only one of the four causes philosophy has known since Aristotle, has set the standard for all causality. So far has modernity carried this that the final cause is no longer even considered a cause but often is dismissed by reason as myth or delusion. Telic finality, or the holy, has disappeared from our thought, been repressed into the shadow.

Moreover, although the theory of causality traces itself back to Aristotle, Greek thought, according to the Thinker, had nothing to do with bringing about and affecting.[7] Instead of imagining the linear movement of atomic billiard balls clicking against each other, a causality modeled on the pure mechanics of motion, the Greeks thought causality as responsibility, indebtedness, and mutual interdependence—a responsibility that, responding to the presence of a calling, draws the thing out of nothingness into being. Causality is not bringing something about but bringing something forth—like giving birth to it, revealing it. The difference is decisive, as big as the difference between fathering and mothering.

For example, the silver that comes from the earth and goes into the chalice as its matter is co-responsible for the chalice.[8] Responding to the worlding of sacred chalices, the meaning of a cup in a holy context, the silveriness of silver gives the chalice its sacred nature, making it special, rare, holy. At the same time, and equally responsible, the chalice is indebted to, and brought forth by, the aspect of a chalice, the cuplike form that marks its difference from a brooch or a ring and makes it useful as a sacred container.

The third and most important part of the chalice is its sacred character, the part that in advance confines and circumscribes the chalice as a sacrificial vessel, the telos of the chalice. But we must not misunderstand telos simply as aim or purpose, as modernity is apt to do, but rather as that which bounds and completes the thing, the world that surrounds it, provides an interpretation and place for it, and makes it whole, holy. The telos of the chalice, its character as a sacred tool, brings forth the silver from the earth and makes it into the shape of the chalice, calling on the craftsman to bring it into the world that worlds it.

The fourth and final participant in the play of co-responsibility that grants the chalice its nature is the silversmith, the one who brings about the sacred chalice. The silversmith is granted her place in the fourfold play of responsibility, not just because she brings about, giving mechanical cause to the effect that ends in the chalice, but because she considers carefully, with the reverence appropriate to her task, interpreting her responsibility to the world that worlds around her, and draws together the silver and the shape of the chalice into its completed whole as sacrificial vessel. It is her handcraft, responding to the worlding of the sacred chalice and interpreting the nature of the earthen silver, along with the functional form that it must enter into, that makes her co-responsible for the chalice. As the dwelling artisan, a mortal drawing near to her time and place and interpreting her responsibilities, she brings the chalice forth from nothingness, making it present in the world, revealing it.

Though the four modes of responsibility are each different from the others, they are all united in the play that brings a thing forth out of nothingness. They set the thing free, gathering it into its place in the world, starting it on to its way of arrival. Every tool that fits the hand of mortals, every tree, rock, flower, everything that appears amid the dwelling place of the mortals, participates in the fourfold unity. The four modes of responsibility bring things forth, revealing that the happening of the thing is indebted to many things beyond itself. It is no mere bringing about but a happening of world.

What, then, is a thing? The Thinker poses the question with childlike simpleness, asking in innocence what reason overlooks because it is too near and obvious. But nothing is murkier than the thinging of things, especially in modern times.

The Old High German word *ding* and the Old English "thing" both mean "a gathering," and specifically a gathering to deliberate on a matter under discussion, a publicly contested matter.[9] "Thing" refers to anything that bears upon humanity, that concerns it as a social gathering, that is a matter for discourse and discussion. The Romans, deriving their usage from the Greek *eirō* (*rhētos, rhētra, rhēma*), called a matter for discourse *res*. Contrary to common knowledge, *res publica* refers not to the state but to that which, concerning everyone, is known by everyone and is debated in public. Often, since the Romans were great legalists, the word *res* designated a case at law.[10] The Romans called this thing a *causa*, or, translated into English, a cause.

We must not, however, understand *causa* in a mechanical sense, but rather as pursuing a cause for justice, a charitable cause, a political cause, and so on. A cause is a thing that is before us, gathers us into its care, and calls on us to be responsible toward it. It rises up of itself out of nothingness, the unknown void, making its presence felt, demanding a response. In such a cause, the world worlds, is revealed. But only because *causa*, almost synonymously with *res*, means the case can the word *causa* later come to mean "cause" as the causality of an effect. The point of pursuing a cause, of responding to the thing at hand, was to produce, shall we say, an effect, a result, in things. Gradually the Roman *causa* became the Spanish *la cosa* and the French *la chose*, or, as we say in English, "the thing."

But, back to the world that pursuing a cause invokes. A thing reveals a cause, a whole world's worlding. Gathering the world together, revealing it, the cause calls its thing forth out of nonbeing, giving birth to it. In the thing's happening, and in the absence that preceded it, that made the cause a calling, the world worlds. The summons that gathers a thing into being, that causes it to be, is, as we saw, composed of four different moments, each one responding to the others.

As a thing, the silver chalice is a gathering together of all that makes it a chalice: the earth, because its dark obscurity is the fertile source of its silver; the sky, because the chalice's aspect and function are revealed by the sky's openness; the gods, because they are the whole, the beckoning messengers calling for the gift of the sacrificial offering; the mortal who makes it, because through her facing her own death, her caring concern for the earth, the sky, and the gods, the gathering of the chalice is drawn forth from unconcealment. Thinging, the thing stays earth and sky, divinities and mortals, bringing the four in their remoteness near to one another, uniting them in the simple onefold of their self-unified fourfold. Gathered together, situated in the thing, the world worlds.[11]

The Thinker uses strange phrases—"the world worlding," "the thing thinging"—in order to draw attention to the situating of things, the be-ing that makes them be. The thinging of the thing, the identification of its presence before the eyes of dwellers, cannot be reduced to reason's causality, a mechanics of stimulus and effect. Instead, it is revealed, drawn forth according to its own mystery from unconcealment, made present and given its identity by the world's worlding. A world has to be dwelled in, a maker has to be situated in body and dress and desire, and the identities of things must be at hand, available for interpretation, before the things revealed in it become present. Otherwise, they are not. There is no independent reality, no truth beyond the world, no way for any-thing to be known without being situated.

The world's worlding cannot be explained by anything outside of it, underlying it, or determining it—by a theory of metaperception, for example. This impossibility is due not to failings of human science or the inadequacies of our reason but to the simple being of the world, the incalculable and unfathomable character of the world's worlding. Modernity's will to explain, to reduce an underlying truth to separate parts and then to capture the structure of cause and effect in a law-governed beyond, strangles the basic nature of the united four. Earth and sky, gods and mortals, cannot be understood by separating them and explaining one by another. They are together. Mortals dwell on the earth, under the sky, in the presence and absence of the gods. Dwelling necessarily brings the four modes of occasioning together and presents it as the thing.[12]

Technology is the *way* this happens, the way the thing things, the world worlds. It can happen in many different ways, depending on the gods that mortals attend to. The thing can happen amid awe and reverence, or amid erotic joy and bliss, or amid mystery and wonder, or amid otherness and difference, or amid power and fear. The chalice that we handle can alternately be a holy grail, a metaphor for the earth's womb, a symbol of the abyss of Being, or a trinket for the tourist industry. Whatever it is, it has little to do with the mechanics of stimulus and response, even if our age thinks quite otherwise. And so technology is not merely a means for man's power, a way of controlling things or bringing them about. That is only how the thing happens in the modern age, as an object of utility and domination. It need not happen that way at all; in fact, it often has not. Technology is a whole world, a way of revealing what things are.

The etymology of the word "technology" is quite revealing.

Technology derives from the Greek word *technē*, which does not mean simply "an art" or "handcraft," as the dictionaries commonly have it, but rather, "to let something appear," "to bring a thing forth from unconcealment,"[13] in short, to reveal it. Technē is a mode of *alētheuein*, of unconcealing, of revealing whatever does not bring itself forth and does not lie before us. Technē refers to the things that humans alone make, the truths that people alone bring forth. As a mode of *alētheuein*, technē is contrasted to *physis*, or those things that come forth of themselves.[14] *Physis* is commonly, but inadequately, translated "nature." Prior to the modern distinction between things humanmade and things that are natural, prior to Aristotle's imposition of a difference between technē and physis, the latter revealed that which bloomed forth of itself, as a rose does. In the time of the pre-Socratics, strangely enough, physis was identical with logos, the language of Being, because then language was not a human instrument but something that came forth of itself from the chaos, the void that was the origin of all things. Physis was the origin of all things natural and humanmade because it was the name of that which, always concealing itself, mysteriously brings things forth. This time did not know a difference between physis and technē.

And it did not yet think *archy*, either. Archy (the common English root, from Greek *archē*), which seems to have come into its own only with Aristotle and Plato, refers to two things: command and origin. Archy is a commanding origin, a cause which calls things into being.[15] As a result of its claim, things happen, but only according to its rule or logic. It is that which governs appearing, appearances. As an underlying and general truth of things, it makes knowledge of things possible by establishing their origins in rule, law, principle, reason, form, and so on.

Once the distinction between technē and physis is made, between humanity and nature, it becomes possible (and only in the modern age is this fully accomplished) to think of Man as an "archy-tect," the one in whose command is the origin that builds things, brings them forth.[16] Since the time of Aristotle, whoever builds—constructs a house, a ship, or a sacrificial chalice or cultivates a vine or whatever—reveals things as they are brought forth by an archy of some sort, a ruling command that gathers together in advance the underlying principle of the ship or the house with a view of the finished thing, and from this gathering rules and governs the manner of its building or construction. The archy is the origin—whether logos, the gods, reason, the laws of nature, or Man himself—of this making of things, not physis, the erupting earth. Things can be unnatural now because they are human-made. And things can be dominated, used as a means, because there is now a difference between humanity and nature. Before, when logos was physis, and everything was one with the earth, such a technical hierarchy between humanity and nature was unthinkable.

Though the Thinker is silent on gender and sex, "man," I think, should be also taken to mean male, strictly male. As Riane Eisler has argued, the character of technology cannot be separated from the relations of the sexes. Before the patriarchal Olympian gods defeated the Goddess in myth and in culture, technology was entirely different. In ancient Minoan Crete, a surprisingly sophisticated civilization where this difference is best reflected, the whole way of life was pervaded by the worship of the Goddess. Descent was matrilineal; women were the true equals with men, if not even their betters. In art women were presented in positions of power—as the leaders of religious rituals, as the ones making decisions of state, and as the ones conducting the affairs of civilization.

Beginning around 6000 B.C.E. and continuing up to the fifteenth century B.C.E., when the culture came under Achaean domination, Minoan culture reached astonishing heights of sophistication, in many ways more impressive than the Greek culture that followed it, if only because of its egalitarian, nonviolent, and nonexploitive character.[17] But it had more than that. For much of its history, all of the urban areas had excellent drainage systems, sanitary installations, and indoor plumbing. There is evidence of viaducts, paved roads, water pipes, and large-scale irrigation works of such quality that few civilizations have ever equaled them. Even the poor segments of society had a quality of life that surpasses that of many in the Third World today.[18]

Archaeologists are uniformly impressed, if not amazed, with the quality of their art, architecture, and social organization, using words like "sensitive," "grace of life," and "love of beauty and nature" to describe it.[19] There is no evidence of warfare on the island, no public expenditures on walls and fortifications, and no glorification of violence, war, or conquest in its art. There is no celebration of authority, no monuments with the names of rulers on them, no identification of power with the ability to humiliate and slaughter the enemy or to oppress and silence dissent.[20] Instead, power in Crete was "primarily equated with the responsibility of motherhood rather than with the exaction of obedience to a male-dominant elite through force or the fear of force."[21] Rather than a means of dominating others, power was an ability to enable, nurture, and cultivate. Expressed in care and mutuality, it was a power with, a power to, instead of a power over.

Technology, instead of a being a tool for man's domination, war, and hierarchy, was a way of revealing the generative, nurturing, and creative powers of nature, from which humanity was indistinguishable. Rather than focusing on the technologies of war, the power to destroy things, the Minoans developed the technologies of nurture and cultivation.[22] In fact, it was Goddess-worshiping cultures that first invented agriculture, as well as the art of making pottery and of weaving and spinning cloth, the phonetic alphabet, not to mention the art of government. (In fact, the credit we give the Greeks for discovering democracy probably belongs to the more egalitarian Goddess-worshiping cultures.)[23] Tending toward union, harmony, peace, equality, and the renewal of the cycles, this way of life never revealed itself as technē, as the power to control nature, or to stand in opposition to it. Unconcerned with its origins or the security of its power to command, it was an anarchycal culture, without the need to subsume things under the authority of an overarching claim.

Building technological archytecture is perhaps a need originating in men's insecurity about their relation to their progeny, about how things are revealed. Women have an immediate, sensual, internal, and unproblematic relation to their children, Dorothy Dinnerstein observes.[24] Men's relations to their children are much more problematic and must be mediated by culture and suprasensual reality because it is external to their bodies. How can they *know* the child is theirs? Only by controlling the revealing of things, by establishing rules for women's sexuality, by defending culture against their passion, and in general by dominating the sensual with the suprasensual.

Men's insecurity, envy, and fear, then, about the revealing of things are the origin of the hier-archys that subordinate the sensible to the intelligible, form to meaning, the visible to the invisible, the maternal bond to the paternal principle.[25] In order to secure their claim on the future, to establish their claim on their progeny, men become the authors of creation, and women are reduced to mere vessels of flesh carrying their seed. Seeking to preserve male author-ity over creation, they establish themselves as the readers of legitimacy and illegitimacy, the unequivocality of meaning, and the certainty of origin.[26] Phallocentrism is thus intimately bound up with patriarchal authority, a technology of domination, and a religion of suprasensuality. While motherhood is like physis, something that blooms forth of itself, like a rose, patriarchy is something that, like technē, has to be brought about, claimed, established, asserted. Patriarchy must create a difference between itself and nature, an archytecture, in order to subject whatever is revealed in it to its reign. Motherhood has no need for such a distinction to claim its truth and thus has no need for archytecture.

The usual conception of truth—the patriarchal conception of truth that abides in our age of reason that the Thinker calls on us to deconstruct—starts with things and ends by establishing their correspondence with a proposition, a universal and eternal archy, such as the will of God (or Man), the laws of nature, the logic of reason, or the mathematics of probability.[27] Truth, as it is commonly thought, is that which correctly corresponds to its referent.[28] Meta-physis-cal in truth, this conception of truth moves beyond the motherly physis the pre-

Socratics knew to the patriarchal archys that we know. Just as the original thought revealed in physis was concealed by the distinction Aristotle made between physis and technē—things that come forth by themselves and things that come forth by the hand of man—metaphysics rests on a sharp distinction between subject and object, knower and known, appearance and reality. As such, it has a decidedly masculine and warlike character to it, bringing a sharp analytic sword to bear on the world, cutting things apart, ignoring the unities that make different things a whole, then subjecting one term to another as its truth.

This metaphysical (and patriarchal) summons for distinction and dominance finds its origin not in Kant or in the modern age with the rise of positivism but in age-old Christianity (Aristotle only opened the way toward it), with the positing of the creator, God the Father, over against the created.[29] The created is true—and humanity is free of sin—when it measures up to, accords with, the eternal and universal idea that God has of it. Man can know the truth of any creation because, created in the image of God the Father, he fits into the unity of the divine plan of creation. His words and thoughts can be brought into perfect accord with God the Father's universal and eternal will because they are wholly adequate to the task.

In our time, despite the absence of the divine in our discourse and despite a popular revolt against traditional metaphysics, the notion of truth as the correspondence between universal and eternal words and contingent things is still retained. Patriarchy still rules. Just as contemporary positivists maintain that scientific theories can be tested and brought into accordance with the facts, Rawls's worldly reason, supplying the law for itself, is able to know the world immediately, to accord the representation with the represented.[30] Empiricists, rationalists, and the holders of any theory that mediates between the two poles remain Christian and united in their notion of truth as correspondence and, despite themselves, remain also metaphysical and patriarchal.

Rupert Sheldrake, who is advancing an anarchycal study of nature that is as persuasive as it is subversive, puts this into sharp perspective. He believes that there are no "laws of nature," only habits of nature, morphic resonances, fields that connect like with like and make them the same. There are no laws of nature because there is no lawgiver of nature; the whole conception is a metaphor "based on an analogy with human laws, which are binding rules of conduct prescribed by authority and extending throughout the realm of the sovereign power."[31] According to him, the metaphor in the seventeenth century was quite conscious and explicit. The laws of nature were legislated by God, the Father of the universe, and his laws were immutable, his writ universal. And if scientists no longer believe in God as lawgiver, they still believe in lawgiving. And that is the problem. These great advocates of truth only with evidence, themselves depend on a method that accepts as given a reality that needs to be proved.

Laws, scientific or otherwise, imply lawgivers, and they imply the authority needed to maintain them. If there is no lawgiver, as most scientists believe there is none of nature, then where is the authority that maintains such laws? Only in the scientific profession. Nature obeys no "laws," speaks with no author-ity—only scientists do. And when they do, they insist upon the most mysterious things. Even though laws have no matter and no motion, they govern everything material and moving. They cannot be seen, weighed, or touched, and yet they are the most important things to be known of nature. They have no physical source or origin that gives birth to any sort of presence, and yet they are present everywhere and always.[32]

The laws of nature are, in short, metaphysical entities, much like spirits, far removed into a transcendental realm that is universal and eternal, there before the universe came into being and there long after it has passed away. If science has killed God, the lawgiver, scientists remain the most devout theologians. Of course the more sophisticated of them would unhesitatingly acknowledge now that the laws of nature have no real or objective existence; they are just theories and hypotheses in human minds. But, dear reader, can you imagine the confusion and the shock in their audience if any scientists ever described their work as discovering the laws of man, or, perhaps, the laws of science? Where would their authority be if they could not appeal to their patriarchal metaphysic? Modern science, lest we forget, is sustained only by the patriarchal power of its metaphors. The claim scientists make on authoritative status is exactly the same as the one theologians have long maintained: the power to make correspondences between the physis and the metaphysis. Only by invoking the holy, a realm they themselves now admit is unreal, do they get the authority to subject the earth to their reading of the beyond.

When truth is understood as correctness, as a correspondence with an eternal and universal principle, or archy, then untruth or error becomes merely incorrect and can be dismissed without hesitation.[33] Falling outside the truth of truth, a shadow that has no reality, untruth is not a part of Being and therefore has no importance once it is understood as untruth. Once metaphysics excludes physis, once the patriarch asserts his claim on revealing, errors are something to be overcome; reflecting no correspondence with the beyond, they have no significance, no presence—in spite of the fact that the world has worlded them, given birth to them. Science, modern scientists will tell us, progresses when we reveal untruth as untruth and then discard it. The illegitimate has no place in their ordering of the world.

But the shadow cannot be denied or repressed like this. It does exist; the world has worlded it. It has happened, been revealed. Though denied and dismissed, discarded as illegitimate, it still belongs with truth. It is what truth is not, and so it gives its identity to truth. The other of truth, beckoning, doubting, it reveals truth as something questionable, something uncertain, something at risk. Untruth is the happening of truth. They are doubles that necessarily complete, support, sustain, and produce each other.

The particular character of the untruth that metaphysics in particular denies, attacks, and represses as shadow determines the whole archytecture of the truth affirmed. Where would modern science be without astrology, New Age mystics, and religion in general? They are superstitious; scientists are not. They believe in the supernatural, which cannot be known; scientists would never make the mistake of basing any claim to truth on something that cannot be experimented with. They are subjective, emotional, dishonest, and manipulative; scientists are objective, detached, rigorous, and bound by the highest professional ethics. They are out for power, to manipulate the gullible, to get money; scientists seek only truth and the welfare of humanity.

Scientists have constituted the others of their truth like this since the Enlightenment, and however much they might deny it, the claim of their illegitimate offspring has secretly shaped the knowledge that they have sought. Scientists who experiment with parapsychology or who offer theories explaining it are automatically cast as suspect, underfunded, and dismissed as loony mystics, even though their science meets the same standards as any of their peers. The same goes for scientists doing work in psychoneuroimmunology, the study of how mental imagery affects bodily health. Subverting the Enlightenment understanding of the body as machine, both of these projects are dismissed and marginalized, not because they are not doing good science, but just because of their proximity to what science is not—that is to say, to metaphysical beliefs, to religious dogma, and to subjective philosophy.

Instead of rejecting truth's shadow the way metaphysics has long done, the Thinker not only acknowledges it but affirms it and celebrates it. The world has worlded it; it belongs and is true to the earth. The otherness of truth, its difference from itself, needs to be accepted as it is, as neither illegitimate nor useless error. Truth depends upon its shadow, is sustained by it, and requires its support. The way to truth is the way to untruth, the concealed, the mysterious. This is unavoidable because, as the world worlds, one thing necessarily conceals another, stands in front of it and perhaps limitless others, obscuring them in unknowable mystery.[34] Truth happens in a particular world, situation, and place, and because it does, it necessarily throws all other worlds into the shadow, makes them into untruth. Because much more is concealed than is revealed when the world worlds, because the darkness that obscures is limitless, while the light that reveals is limited to the clearing where the thing things, the concealment of things as a whole is more primal than any temporal revealing of them. At the edges of the clearing where the world worlds, surrounding it in an endless night extending without limit, time, or direction, lies the mystery, a pure expanse of nothingness. Not a particular mystery, regarding this or that thing, but a mystery that is one because it cannot be divided.

In dwelling, in drawing things forth from unconcealment into a time and place, as the world's worlding calls on them to, mortals also conserve the mystery, preserving the shadow that gives birth to the truth. To dwell, and to think, is to call upon the shadow, to preserve the mystery while revealing time and place. Interestingly enough, according to the *Oxford English Dictionary*, the word "dwelling," besides meaning to stay in one place and preserve it, also means to lead into error, to retard and delay.

By letting the shadow be, by dwelling amid untruth and mystery, mortals free themselves of patriarchy's archytecture of exclusion and gain their possibility for truth. For the Thinker, freedom, which he describes as the preserving and sparing of the thing as it rises up of itself out of the void, is the happening of truth, the revealing of all knowledge.[35] Truth happens only because freedom has already happened. Placing the happening of truth in freedom, in letting the mystery be, does not plunge it into the arbitrariness of human caprice, the wandering subjectivity of an autonomous human will, because freedom is not something humans possess. Freedom is a way the world worlds; it has nothing to do with human choice or will or consent, either individually or collectively. Freedom has nothing to do with human choice because the mystery that calls truth into the world is the shadow cast by the world's worlding, and that cannot have anything to do with will or choice because it precedes them.

The patriarch's metaphysic of truth, seeking to establish its claim on revealing, celebrates the true and dismisses untruth.[36] The Thinker, letting the world world, is grateful to both, acknowledges both because they reveal earth, the happening of things. A dark concealing mystery, the earth is the abyss on which the world is, the calling that reveals truth, setting the world free. The earth, as the Thinker knows it in his later thought, is identical with what the pre-Socratics knew as physis. It is the concealing mystery, the abyss that gives birth to the world, the shadow from which things are revealed. From its dark obscurity things are always coming forward, being revealed, suddenly and mysteriously appearing as truth.[37]

Since the earth is more primal than it is, humanity is not at all in possession of the earth. Our reason, whatever our patriarchs insist upon, does not posit or form the things that appear from the earth, but rather it itself is formed within the earth's dark mystery. As the pre-Socratics knew, logos is physis. Contrary to reason, untruth or error is not founded in the finitude of Man, but in the earth itself. It is the earth breaking forward, revealing the world that it bore within itself, that grants Man the truth he knows and the errors he does not. And so it is not Man that interprets, thinks, speaks, and acts but the earth itself.[38] Our thoughts think us, not we them.

Far from being located in a human will, freedom is the world's happening, a way of Being that lets things be the things they are, all according to their time and place. However, this is not a justification for fatalism, for accepting injustice, exploitation, violence, and destruction. On the contrary.[39] Far from being a passive response to the world, a justification of the negligence and indifference toward things that finds its conclusion in the destruction of the earth and the depravity of modern work, freedom is the call to spare and preserve, to nurture and cultivate. Only by letting friendship and well-being happen does truth happen. Being in the world is a call not only to be with others and care for them but to nurture the whole world.

As Fred Dallmayr has argued, "The genuine or unperverted exercise of freedom is shown to be a persistent tendency or inclination toward the good life, that is, toward human reconciliation and peace."[40] By living in solidarity with other people (and, Dallmayr neglects to mention, the whole earth), by being *friendly* toward the world, truth happens, and the earth is set free. This good life cannot happen alone but only with others because only with their cooperation and help can truth, as friendship, happen. Friendship needs a place to be expressed, felt, realized. Only when such a place is opened up are we free.

A public space, as Hannah Arendt has argued too, is essential, because it is there that we become aware of freedom—and its opposite. Freedom happens in a public space not through the expression, development, and realization of choice or sovereignty but through the happening of truth. In fact, freedom and sovereignty are not identical but opposites and cannot even exist simultaneously. "Where men wish to be sovereign, as individuals or as organized groups, they must submit to the oppression of the will, be this the individual will with which I force myself, or the 'general will' of an organized group. If men wish to be free, it is precisely sovereignty they must renounce."[41] Freedom is not the possibility of choice, the exercise of sovereignty, the assertion of the will; rather, it is the way people are with each other and the whole universe, it is the way they let it happen. They can either let it happen in a free way—that is, in friendship, poetry, and song—or they can try to control it. Truth can happen in friendship; it is impossible with control.

In order to dwell free, humanity must come to know how the archytecture that its patriarchs have built with will and domination conceals and seals off the anarchy of the earth.[42] It must recognize that patriarchal metaphysics cuts it off from the physis, the earth on which it dwells. To live free, humanity must build not according to universal and eternal truths but according to the dispersed and anarchycal truth that the earth brings forth in its time and place. Truth must cease to be universal and eternal and become temporal and local, cease drawing everything around the orbit of a summoning center and let chaos be. It must, in short, open itself up to what presents itself. Anarchy must replace patriarchy's totalitarian claim of being the originary archy.

It is through wandering ways of thinking that humanity gains freedom, accepts its errors, breaks free of its archy, deconstructs its centers, and attains the truth of its mystery. Being unafraid to encounter mystery, accepting its shadows, never forgetting its errancy, thinking freely lets the anarchy of the earth break forward, lets Being be, lets the world world. In this, thinking stands opposed to mere reason, which, clinging to its purity, denies its shadows.

Tracing inconspicuous furrows in language, thinking is a friendly openness that does not disrupt the concealing mystery, the earth that makes all things possible, but draws it unbroken into the openness of understanding and truth. But while thinking regards the world with gentleness, it is not necessarily kind to the sophistry of reason.[43] Standing opposed to reason, thinking must, as the breaking forward of the earth, be discordant and disruptive, letting strange and unfamiliar thoughts rise up against the organization of reason and common sense.

To be free is not to be at the origin of one's actions, the rational self-willing cause of all that one is, but rather to let the world world, the thing thing. As Fred Dallmayr has argued, even though the Thinker has refused to develop an explicit political theory, this conception of freedom does have definite implications for political action.[44] Subverting subjectivity, modernist notions of causality, and the correspondence theory of truth, he subverts the archytecture of almost all our institutions. In the modern age, institutions ranging from the scientific to the educational, the political, and the family are built around the understanding in some way that people are free when they can control their environment, whether social, personal, or natural, and make it submit to their will.

Following the Thinker's reading of the modern age, there are many ironies built into our various institutions' quest for freedom. We can state the irony this way: if freedom is control, then control is freedom. To be free, we must be able to control every aspect of ourselves that we want to free. In other words, we become authors of our life only to the extent that we are controlled by it, to the extent that our thoughts, actions, beliefs, relationships, and whatever are all controllable, subject to causal manipulation. Modern freedom is very demanding. It has a lot of disciplines shadowing it, a lot of institutions built around protecting it. People must be made rational, responsible, orderly, law abiding, cooperative, normal, and so on. So they are put in institutions that produce these qualities—the school, the military, the factory, the government. And each one of these disciplines authorizes its shadow, people who fail these disciplines. They are read as irrational, irresponsible, disorderly, delinquent, uncooperative, subversive, and perverted; they are housed in institutions for the mad, the criminal, the deformed, and the politically incorrect—those out-of-control people.

To get control over their lives, people must be disciplined, forced to be free, or else freedom—that is to say, control—will not be theirs. If people are not disciplined, they are not available for production or for us to control, and so we are all less free, less able to control our world. Perhaps it is because our freedom depends upon *their* submission, our control upon their being controllable.

Political elections in America, according to liberal political theory, are supposedly the means by which our government re-presents individual choice, makes the government subject to popular will, makes the state legitimate and sovereign. But in practice, thanks to modern technology, elections reveal something entirely different from individual choice. Deploying the very latest in political technology, the candidates compete against each other by means of fund drives, public polls, voter analyses, political commercials, and professionally written stump speeches. They struggle with each other to exploit class resentments, sexual and economic insecurities, racial hostilities, and chauvinistic sentiments in the public's mind, and then they convert the hatred, fear, or resentment that they have unleashed into political support. Issues are debated not to explore all their implications, to make an informed decision, to enlighten the public, or, especially, to reveal truth; the aim rather is to invoke symbols and prejudices, inflame emotions, and caricature opponents. When they are over with, American elections establish nothing so clearly as which candidate has the greater power to read, manipulate, and control the public. No one can say that truth has in any sense happened, except the truth that we are dominated by something far removed from ourselves.

Elections are not the only institutions that truth does not happen in. So that it can turn more control over to the individual and promote consumer sovereignty in the marketplace, the liberal state draws a sharp line between private and public, the government and the economy, and lets the market be free. That freedom of the marketplace is privately controlled, assumed for the most part by the multinational corporations. And the elites of these corporations are not bashful about asserting their discipline or their power. At the slightest whim, accountable to no one but themselves, the corporate authorities will lay off tens of thousands of workers, speed up assembly lines, displace workers with new technology, move factories overseas, pollute the workplace or the environment with deadly chemicals, or defraud the consumer with dangerous or useless products. And then they call this "free enterprise." Of course it is—to them. They control the whole production process. And of course they rise up with indignation as soon as someone would limit their authority over their assets, their workers, their consumers, or their environment. That would, shades of Marxism, be a limitation of their freedom, their right to read the whole world as their utility. And they would be right. Freedom depends on control.

Unhappily, none of this would be much improved or significantly changed if somehow the control of production were socialized and then democratized, for that would only universalize the tyranny of control. It would not free anyone from the demands of control but would only make control a universal possession. Even if freedom's disciplines are something we choose for ourselves, something we reflectively consent to and will, they still assert their demand on us, subjecting us to their logic, their imperatives, and their truths. If we are free, then we are controlled. Seeking freedom this way, the Thinker would argue, we are not free. We cannot be because we are seeking control.

Freedom, thought in its most profound way, is not control but friendship, not authority or hierarchy or patriarchy, but thinking in a way that lets differences be. Freedom is a friendly way of being, the way true friends are with each other—sparing, caring, loving. They leave each other be, free of control, manipulation, dishonesty—free to become the best they can be. Freedom as control, whether liberal or Marxist, is a profoundly unfriendly way of being. Seeking to control things, they both set upon them, challenging them to become something else than they are, demanding that they become useful to something distant to themselves. Instead of letting them find their own meaning, they impose the archytecture of their unqualified way. Everything becomes a means to this freedom, turning everything—people, animals, nature—into its slave and thus its enemy. For how can one be a friend to anything one controls? The only relationship that freedom as control can know is suggested by the words struggle, war, conquest, and domination, as well as manipulation, deception, submission, subjugation, and exploitation. There is nothing friendly about freedom as control, even when it happens as democracy and voting, and so there is nothing free about it. Only enemies, opponents, victims, and victors glare at each other in this world of control. No truth can happen in that kind of place.

For truth to happen, the world must world in a way that cultivates inherent worth, equality, difference, assured respect, careful listening, thoughtful meditation, and peaceful engagement. Fear, hierarchy, resentment of domination, inequality, metaphysical judgment, the possibility of rejection, exclusion, and humiliation all work against the happening of truth. It is not enough to give people control over their lives, as some forms of democracy attempt to do. Much more than that, they must be friendly and peacefully engaged, not only to each other, but to the whole world. The birds in the sky, the fish of the deep, the animals of the land, the mountains in the mist, the stars above: all must be listened to, respected, and spared.

Though there is a danger of idealizing them, American Indians lived this way. One of the political institutions that they gave the world is the caucus (the word comes from the Algonquian languages). In a caucus people talk with each other in a friendly way without ever voting until a consensus is reached. It often lasts a long time because it is important for everyone to be heard, to work out their problems, to feel sure that all of their concerns are respected. In comparison with even democratic procedures, caucuses are less divisive and combative because all the participants are assured of their worth and do not risk humiliation.[45]

The Indians treated the rest of the earth the way they treated each other in caucuses. Animals were talked to and taken into account, as were the mountains, the trees, the waters, and everything else. A group of deer, according to one old man in a novel by James Welch, talked to him "about days gone by." They were not happy with the way things are. "They know what a bad time it is. They can tell by the moon when the world is cockeyed."[46] Gary Snyder, a contemporary poet, advocates a kind of ultimate democracy that includes the nonhuman, the other. Plants and animals are given place and voice in this democracy through rituals and dances, and perhaps through shamans who become them then return with their understanding of the world.[47] Many modern humans, raised by the claims of science and patriarchy, may find it strange to include all the earth in their community, may wonder about the legitimacy of a state that included all the plants and animals in its decisions; perhaps until we do this, however, truth will not happen.

CHAPTER THREE

Science *and* Technology

> The ancient masters of science promised impossibilities and performed nothing. The modern masters promise very little; they know that metals cannot be transmuted and that the elixir of life is a chimera. But these philosophers, whose hands seem only made to dabble in dirt, and their eyes to pore over the microscope or crucible, have indeed performed miracles. They penetrate into the recesses of nature and show how she works in her hiding-places. They ascend into the heavens; they have discovered how the blood circulates, and the nature of the air we breathe. They have acquired new and almost unlimited powers; they can command the thunders of heaven, mimic the earthquake, and even mock the invisible world with its own shadows.
>
> —Frankenstein's professor, in Mary Shelley's *Frankenstein*

IT WOULD SEEM that modern technology, unlike ancient technology, is based upon, depends on, and is the consequent of the exact mathematics of modern science, the practical application of what science is in theory. Ancient technology, it is true, did not rely on science and mathematics to do what it did, and so it could be a handicraft, the concrete application of everyday knowledge. But the Thinker maintains that this understanding modern science and technology have of their relationship should be reversed.[1] It is not science that makes technology possible; it is the technical ordering of the world, its archytecture, that makes everything over into man's utility, that makes modern science possible. Technology is the way the modern world worlds, and mathematics is only a means in that way.

Besides, most new discoveries are made, not because they have been anticipated by science, but because someone has perfected or developed the experimental apparatus that presents the data either in a different, more precise way, or because someone has developed a new technique or procedure for dealing with the data. As Bob Ackermann and others have argued, instruments, or technical apparatus, are an essential moment in a dialectic that moves between data, instruments, and theory.[2] For instance, new data domains are uncovered as the techniques for producing them are perfected and developed. A slight improvement in technique will sometimes produce anomalous facts, previously ignored because the technique that made them was not judged sophisticated enough to explain the abnormality. These new facts cannot be explained by prevailing theory but can only be understood by a different theory. It is the craftsmanship, the technical character, of science that guides its development and determines its truth.

According to the Thinker, "science," as we understand it today, is fundamentally different from the *doctrina* and *scientia* of the Middle Ages, and especially from the Greek *epistēmē*.[3] Unlike modern science, Greek science was not exact in calculation or measurement, nor did it need to be to respond to the things present to it. We cannot claim, then, that our science is more true because it is more exact, more mathematical, than theirs. Nor can we say Galileo's doctrine of free-falling bodies is true and Aristotle's teaching that light bodies strive upward is false. That is like saying that Shakespeare's poetry is more advanced than that of Aeschylus's. Because of the different technologies, the different ways the world worlds in different times, we cannot say that our science is more correct than another time's or that it has progressed beyond it. Between the world of the Greeks and our world there lies a break, a gap of incommensurable understanding, that cannot be bridged by continuities of measurement and development.[4]

Even if it were, it is hard to know what should be measured. It may well be that in terms of fulfilling emotional needs and promoting psychic well-being (hardly an irrelevant consideration), the ancient Minoan civilization was far in advance of our own. They seem to have found a joy and meaning in life and a tranquil acceptance of death and pain, according to those who have studied them, that totally eludes all modern civilizations. It was a social order in which, writes Nicolas Platon, "the fear of death was almost obliterated by the ubiquitous joy of living."[5] And there are existing today "primitive" cultures that do not experience mental disease of any sort. Thus, to understand the nature of modern science and its technology, we must free ourselves from the tired habit of comparing the new science with the old solely in terms of degree and progress, of judging the past in terms of an archytecture that only modern reason knows.

Science is technology, a way of knowing things, because it depends on the scientific method, as we saw in the last chapter, for establishing the truth of its claims. Things are true for science because the correct method—that is, the politically correct technology—has been followed. Every technique or procedure of knowledge, by the fact that its own way is followed, determines beforehand that which is, the world that worlds there.[6] Through the ruling governance of the world's worlding, it uncovers the things—natural events, reactions, processes—appropriate to it and locates them in their unique ordering. And it casts everything else into the shadow, either actively through repression, or unknowingly through concealment.

For example, modern physics, the paradigm science of modernity, uses mathematics obsessively. But it can use mathematical technique only because it already is mathematical.[7] In its world, things are present and understood exclusively in terms of its mathematics. This is quite in contrast to the Greeks, who, while being excellent mathematicians and who included the great master Euclid among their number, did not understand physics as numerical or calculative mathematics. Aristotle, for example, did not use numbers or equations to explain the physics of things.

For the Greeks, *ta mathēmata* meant that which Man knows in advance in his observation of whatever is and in his relation with things, whether it was the corporeality of bodies, the vegetable character of plants, the animality of animals, the humanness of man, or the exactness of numbers. Just because numbers are the most familiar and striking of the already known does not mean that originally the mathematical is totally encompassed by numbers and calculating. On the contrary. As the already known, the mathematical had a much more expansive domain, covering many more qualities and properties than are contained in numerically exact formal systems. That experience of mathematics has been lost in the shadow that modern technology casts.

When Galileo was developing his idea of inertia, he did not passively observe the movement of things as they naturally occurred, as Aristotle would have, but rather he set up an artificial idea, a hypothetical universe that existed only in the mind.[8] By means of this utopia, which removed such variables as air resistance and friction from the experiment, he reduced the pure inertial movement of things to a mathematical formula. Creating a placeless place in his mind, he stipulated a numerical world that became the perspective from which all of nature was to be plotted out and understood. Despite nothing in nature behaving exactly as the formulas predicted, despite their literal unreality, Galileo's mathematics have overwhelmingly been accepted as true, at least in nonrelativistic normal circumstances, by generations of scientists ever since. The mathematical projection, out of its own ruling, provides its own standards of exactness.[9] And now the world is governed by the utopian expectation of mathematics, without appeal to any other standard. It is indeed ironic that Aristotle, a far less utopian thinker than any modern scientist, has for centuries been vilified for being a poor empiricist.[10]

In general, physics is the knowledge of material bodies in motion. And if the archy that governs physics is calculable, if everything that it encounters—every motion, every reaction, every weight—is quantified, measured, and subjected to calculation, then that is what nature becomes, namely, something mathematical, numerical, and formalizable. Motion becomes a measurable change of place, and no motion or direction of motion is superior to any other. Nor is any place, in this placeless utopia, superior to any other. All forces are defined according to their consequences, their quantifiable, measurable effects of transformation. Causality becomes understood as a mere bringing about rather than a bringing forth. All events are only inasmuch as they fit within this utopian projection; outside it they are not. This projected ordering of nature into the mathematical finds its guarantee in every technique applied to the body of nature. What will be learned is already known by its mathematics, and its value is determined by its measure and formal exactness.

The mathematical projection thus makes possible our mechanized and rationalized world by reducing the fourfold modes of occasioning to mechanical causality, a simple linkage of material cause and effect. In order for it to attain the rigor that it demands of itself, for it to place itself in its utopia, mathematical reasoning must remain within the framework of its initial premises, drawing upon no assumptions, tacit or otherwise, that were not explicitly postulated at the outset. Modern mathematics is to the core of its nature axiomatic, a formalizable procedure played out within explicit and unambiguous rules of organization. Abstracting itself from all external meaning, seeking to eliminate any appeal to evidence that rests on intuition, modern mathematics becomes nothing but an internally organized and tightly specified configuration of symbols and procedures.[11] These symbols and procedures are so highly organized and tightly specified that a machine, if it were sophisticated enough and properly programmed, could perform all the operations of mathematics automatically.

From the beginning it was as if the whole necessity of modern mathematics was to advance itself to the point where it became a gigantic, infinitely exacting machine that could automatically grind out all the truths of mathematics, completely eliminating the creative role of the mathematician. It is no accident that our age has created the computer. It was a machine made necessary, if at first only in metaphor, by the destiny of mathematics itself.

As it is with a mechanic who is creating a well-running machine, the task of the modern mathematician is to create a complete system without any contradictions in its axioms. (This is a generalization, I must admit. Some modern mathematical systems, like those developed by Kurt Gödel, are unavoidably inconsistent and incomplete. Mechanical procedures break down in it. But this anomaly does nothing to change the character of the reign of mathematics in our world.)[12] Because the truth of mathematics rests on its own internal structure and not on any form of external evidence or intuition, the truth of any one conclusion, however small or insignificant, is totally dependent on the integrity of the whole structure. The truth of mathematics thus depends solely on its consistency, its mastery of contradiction and ambiguity, its total repression of its shadow. A machine is useless, worn out, or broken if the interactions of its assembly do not by themselves produce the desired result. When we wash our clothes in an automatic washing machine, we expect the machine to go through a decision procedure, filling up, agitating, spinning, rinsing, agitating, and spinning, all by itself. As with the washing machine, mathematical procedure, following the axioms and definitions imprinted within it, must produce unvarying results, universal truths that any mathematician would come to if the correct procedure was followed. Modern mathematics is pure technique, a total reduction to mechanical form and causality.

But the rigorous demands of exactness in physics as mathematical research, the Thinker tells us, are not simply reducible to quantified precision. It seeks quantified precision only because the things present to its reasoning demand mathematical rigor.[13] If the objects of research cannot be measured with quantified precision, as they cannot be in the life sciences, the human sciences, or the historical sciences, then the rigor of mathematical research is satisfied with an analytic and formal discourse. The lack of quantification in these sciences is not a deficiency but only the appropriate response to the formal nature of this kind of research.[14]

In all these sciences—physics, as well as the life, human, and historical sciences—science becomes mathematical research through the projected plan and through the securing of the plan in the rigor of rational procedure, the technology of accumulating knowledge. If the world mathematical exactness projects is to become secure, objective in its truth, then it must be able to encompass all the facts that exactness uncovers under the rule of laws. For only within the ordering of rule and law, the techniques of mathematical procedure, do facts become clear as the facts that they are. Facts in all their plenitude and diversity are mediated and made possible only by the rules, laws, and instruments of measurement that anticipate them.[15]

Clarifying itself on the basis of what is already rational to itself, mathematical explanation is the encompassing of facts under rules and laws. As such, it is always twofold in its nature. It accounts for an unknown by means of a known, and at the same time it verifies the known by means of the unknown. A new technique of knowing unique to modern science—the experiment—is what modern science uses to explain the known and uncover the unknown.[16] The experiment is possible only after the knowledge of nature has been transformed into a calculable presence and science is understood as mathematical explanation.

Although they were often very careful in their observation of things, and demanded compelling evidence for the assertion of any claim, neither medieval *doctrina* nor Greek *epistēmē* performed experiments as modern science does. Contrary to common belief, the Thinker argues, Aristotle was a very careful observer, observing things as they presented themselves, their qualities, their modifications under changing circumstances, and so how things as a rule behaved. He was a very good "empiricist," describing the limits and possibilities of observation, the circumstances under which it was distorted or clear. But Aristotle's "empiricism" remains fundamentally different from the modern experiment, and it remains so even when ancient and medieval observation works with number and measure, or uses apparatus and instruments.

Despite seeming similarities that have been interpreted as the origin of the modern experiment in ancient knowledge, a decisive difference remains, separating their techniques of knowledge from ours. "Twisting the lion's tail," as Francis Bacon would say, the modern experiment disturbs things, separates them from their context and subjects them to rigorously planned control to reveal the mechanical causes determining them. Invoking this utopian perspective, it begins by laying down a law as a basis for the experiment, abstracting the thing from the forces, variables, or complications that make impossible a formal observation of the facts or any possibility of the determining causes that will either prove or disprove the law. Where Aristotle sought to observe things in their natural condition, surrounded by all the four modes of occasioning in all their complexity and interplay so that he could know the thing as it is, the modern experiment, invoking its utopian world, establishes artificial, controlled, and planned circumstances, eliminating the complexity and interplay of "irrelevant" variables based on an already-known law so that it can know the simple mechanical causes of the thing. This controlling based on an already known, this reduction to mechanical procedure, makes the truth of the modern experiment mathematical and distinguishes it from ancient and medieval observation.

Contrary to the experiment's mythology, its self-proclaimed progress is not a progress that all ages could acknowledge and revere. By abstracting itself from the natural situation of the thing, by reducing its truth to a mechanical cause or a mathematically formalizable entity, the experiment conceals many things. For instance, the use modern agribusiness makes of chemicals to kill directly the pests that damage crops could have arisen and been acknowledged as an improvement of technique over the older methods only because the experiment concealed the true complexity of nature from itself. Aristotle, for instance, would not have been convinced that the development of a chemical that kills a pest simply and directly would be a useful tool, an advance beyond old techniques, because, by abstracting the insect from its environment, the experiment ignores the place of the insect in the nature of things, making its environment vulnerable to external consequences that it could not anticipate. What other insects would the chemical kill? What effects would that have on other life-forms? What would it do to the soil after many years of use? What would it do to the farmer, her state, her gods? In order for a chemical pesticide to be judged an improvement for Aristotelian science, it would have to open itself up to many more questions than it does in modern experimental science. An Aristotelian science might well reveal that organic farming is a much better technique than chemical farming because it looks at a technique from the vantage point of its place in the whole of things. Chemical farming can be revealed as an improvement or as progress only by invoking the utopianism, the unsituated perspective, of modern science.

It must, however, be cautioned that the already known, the projected law that controls the experiment, is not an arbitrary imagining. It comes from the presencing of things drawn out of nature and present to the plan of the experiment. The experiment is a technique that, in its projected plan and execution, is founded on previous experiments and the things made present in them. And the more exact the projection of nature and the things revealed in it, the more exact becomes the possibility of the experiment. The experiment is a technique of knowledge that spirals inward, ever more exactly establishing the conditions and truth of its own knowledge.

Although the social sciences do not always attempt to trace facts back to laws and rules, they never limit themselves to merely reporting the facts. Just as in the physical sciences, the social sciences aim to fix, objectify, and render stable the object of their discourse—human beings.[17] And if this process does not always yield universal laws, it never fails to compare everything with everything, explaining everything against the ground plan of history, subjecting human life to the calculation of cause and effect, to the measure of norms and equivalences, the ordinary and the average. The unique, the rare, and the great are eclipsed and rendered invisible by being incomprehensible in the presence of the norm.

As science as experimental research progresses, building on itself through ever more exact experimentation, it relies increasingly on itself, turning inward and relying on proven techniques, methods, and procedures. A fact gains currency and value only because it arose out of an accepted technique, a method that a community or institution of scientists believes will produce truth.[18] Because truth arises out of correct method, that is to say, an institutionally approved method, science as experimental research is of necessity institutional research.[19] Results from experiments must be precisely detailed, communicated to other scientists, and reproducible by other scientists. All this takes a closed community of specialists who write articles for one another, talk with one another, and meet one another at conventions. Based on approved techniques, these institutions develop bureaucracies for the funding of experimental apparatus and for selecting those who will receive funding and those who will not, as well as bureaucracies to decide whose results will be published and whose will not. Science as experimental research produces a regime of truth enforced through disciplinary bureaucracies of correct method.[20] Dissent from the established reign of truth is not treated nicely. Tenure may be denied, funding for experiments may disappear, professional stature may be eroded, and the victim may be stuck with all the psychic burdens of being a "failure" and an outsider. As a result of this "professionalism," dissident ideas are often quickly banished to the shadows for reasons that have nothing to do with "good" science. And the reign of orthodox truth continues, secure from threats against it.

As the institutional character of scientific research extends and consolidates its reign of truth, it more and more establishes the priority of method and technique over whatever is. Facts are true not because they are, not because they rise up from the earth and present themselves in their temporality, but because they are established by an institutionally accepted technique or method. Truth comes not from things thinging, from the breaking forward of the earth, from attending to the whisper of the world worlding, but from a regime of truth that conquers and subjects everything to the reign of its method, its reason, and its technology.[21] Before this regime of truth that is science as experimental research, everything is penetrated by rational method and made objective, and the gods whose presence formerly granted things their boundary, their interpretation and place in the order of things, and their sacred character are put to flight.

Through the institutions it calls into being, the techniques it uses to validate its truth, and the rigor of its method, science as experimental research disciplines and creates a new kind of knower—the researcher. The erudite scholar as well as the thinker and prophet disappear, silenced by the institutions of experimental science. The researcher who takes their place no longer needs a library at home, a meditative place for thinking and reading, and certainly not a desert to retreat to. Instead of seeking a silence to nurture the mystery, a place within herself to acknowledge her own shadow, the researcher is constantly on the move, negotiating at meetings of peers, collecting information at conventions, administrating and participating in the bureaucracies of the experiment.[22]

As a result, the researcher is quite unaware of herself and of the way that her own inner life is projected onto the world she studies. The ancient alchemists, who are often dismissed by the modern researcher for their wasted pursuits, were quite aware of how their inner life was transformed by their outer experiments on nature. Writes Paracelsus: "When a man undertakes to create something, he establishes a new heaven, as it were, and from it the work that he desires to create flows into him."[23] The way the outer is treated is the way the inner is treated, and if the researcher sees in nature only something to be experimented with, something to be used, manipulated, and measured, she will cast into her shadow all the parts of herself that cannot be used, manipulated, and measured. The alchemists approached their experiments in prayer, awe, and reverence because they knew what modern scientists have denied themselves, that they would become their work. Alchemy was not so much an effort to dominate nature, to extract gold from straw, as it was to transform the soul. Making gold out of straw was only a metaphor for them for making themselves into gold.

Science as experimental research is able to forget what the alchemists knew because it produces and constitutes the thing as an unconnected object, something distinct and separate from the human subject as knower. Cutting knower off from known, seeking to control the object of knowledge as much as possible, believing the known has no relation to the knower other than the one they impose on it, modern experimental research and the researchers who pursue it can have only a relation of domination to the world. All other possibilities are cast into the shadow.

Cut off from the object in every way except the experiment, the subject as knower gains knowledge of the object through representation, re-presenting the thing in the calculating projections of science as experimental research.[24] Nature, by being calculated in advance as scientific object, and humanity, by being elevated to the status of scientific subject, are "set in place," produced as the doubles of a way of Being that makes possible a rational explanation of everything.[25] Only insofar as the things of these sciences become objects of representation for subjects can they exist at all. In the world of science as experimental research, things simply are not unless they are as formal representations posited by subjects. Only as representation, secured by the technology of the experiment, does science as research attain certainty of its truth.

Thus, for the Thinker, Descartes prepared the metaphysics—even Nietzsche is included in it—for the whole of the modern world, a metaphysics that in its relentless quest for objectivity, its production of the subject, and its use of the experiment as a value-neutral means for knowing the truth of things, is unique among all the ages of the world.[26]

The unique thing about the modern age is not just that humanity has freed itself from the obligations and limits of the Middle Ages but that the very truth of humanity itself has changed—humanity has become subject, or again, Man, in a way never before possible. The Thinker understands the subject as the Greeks thought it, as *hypokeimenon*, as that-which-lies-before, gathering as ground or foundation everything into itself.[27] Originally the Greek subject lacked any special relationship to humanity and none at all to the I.

However, when, as in the metaphysics of Descartes, Man becomes the archytect of the world, the underlying reality of all that is, Man either individually or, later in Marxism, collectively becomes subject. When Man becomes subject, the comprehension of what is as a whole changes, breaking radically with the worlds of all previous ages. The world becomes understood as picture, as re-presentation of things gathered into objectivity and set in place before a human subject.[28] To understand the world as picture is not just to see it as picture, an imitation of the thing, but, much more, to be involved with it as picture, to dwell in a world of pictures. In the world as picture everything in it becomes present only as representation, an archytecture built by Man.

As an example, consider a home-video camera and screen, a technical device of our time that has gained widespread use. Even though some of us might have momentary difficulties making it work, almost everyone can run it eventually, though after some effort and usually after reading the instructions. And everyone understands what they are seeing when they watch it and what the image it presents is. They see themselves if the movie is of them, their friends if the movie is of their friends. Anthropologists, however, once came across a hunter-gatherer tribe that after being shown a home video of themselves reported seeing nothing but shadows and flashes of light on the screen, even though the anthropologists clearly saw pictures of the tribe going about its everyday activities.[29] The tribe members actually could not see the representations of themselves, no matter how hard the anthropologists pressed them.

Having never experienced the world as picture, as re-presentable object, the tribal members could not experience any similarity or difference, any identity, between themselves and the representations of themselves. To them the world was an immediate temporal experience, in no way present as re-presentation. A re-presentation of their world was an impossible thing for them, an identity as equivalent of unequivalent things. They did not dwell amid the tools of representation or amid the identities between things that determine their use. And the thinging of the thing, the home video, simply could not occur for them, become present in their world. They were not gathered into the fourfold occurrence of our world as picture.

Perhaps not all to their loss. This same tribe could, the anthropologists observed with more than a little awe, travel through hundreds of miles of rain forest unerringly without a compass. Any Westerner would only have seen identical trees and would have been hopelessly lost. But the hunter-gatherers saw differences—perhaps differences that no Western eye could ever see—that enabled them to see where to go.

The thing that is present to us, the home video, cannot be reduced to a pure mechanics of causality. Although the physics of perception, the photons that resulted in sense perception, were no doubt the same for both the anthropologists and the members of the tribe, they did not cause the thing, the re-presentation of tribal life, to present itself to the members of the tribe.

Since the world has become picture it has become possible to contrast a new age with the preceding one, to conceive of different "worldviews."[30] In the Middle Ages a different worldview was unthinkable.[31] For that which is, is created by God, the highest cause. To be is to be put into God's created order, to correspond to the will of God. And nothing is unless God wills it. The world of the Middle Ages could not conceive of things being placed before Man's knowing and always at his disposal. That would be usurping God's place, a sin of pride.

Even less could the Greeks have conceived of Man as subject or thought the world as picture. Parmenides said, as the Thinker translates it: "The apprehending of whatever is belongs to Being because it is demanded and determined by Being."[32] Whatever is, thus, is because it arises and breaks forth from unconcealment, coming to presence before Man as the one who opens himself up and comes upon the thing. What is, comes to be not because Man looks upon it, representing it to himself as a subjective perception; rather, what is, comes to be because humanity is looked upon by that which is. To be human is to be beheld by the earth, to be drawn to the thing and borne along by it, even if it is to be driven about by its oppositions and marked by its discord.

The early Greeks were called upon to gather, to save, and to preserve whatever came upon them, even sundering confusion. Because of their calling to preserve even the chaos of the world, the early Greeks could not live the world as picture. It would have required that everything be subjected to a rational ordering posited by Man, throwing chaos into the shadow. Nevertheless, the Thinker points out that when Plato, radically separating appearance and reality, defined the appearance of whatever is as *eidos*, "aspect" or "view," his thought, often ruling indirectly and in concealment, destined far in advance the world's becoming Man's picture.

In contrast to the pre-Socratic mode of knowing things, modern representing brings what is present at hand before Man, forcing it into an exclusive relation with humanity. It is only through Man, his will to power, his mode of production, his sense data, his intersubjective consciousness, that things are. For the first time in the ages of humanity upon the earth, Man decisively and expressly sets himself up as archytect of the world, narcissistically making himself the origin and ruling archy for everything. And so, it is only through Man's mastering himself as a species being for itself, or setting aside the religious delusions of his superego, or becoming the overman, or whatever, that he fulfills his truth.[33] For the first time humanity, as Man, comes into being.[34] And his calling from out of himself is to become master and archy-ical lord over all the earth, to subject to control all that has eluded him in his alienated consciousness, his unconscious delusions, his lies that he has told himself in his weakness. Separated from himself by age-old delusions, Man must come to himself by seeing all that is as only himself, master and underlying truth of the world.

Only because Man has become subject does it become necessary for him to ask of himself the specific nature of his mastery over the earth. Is it as an "I" contained by its own preferences and freed by its own arbitrary choosing that Man becomes subject, or does it happen as a "we," a collective self through which each sees itself as a whole and attains its freedom? Is it as individual or community, as unique personality or mere group member in the corporate body, as nation-state, or race, or world-united humanity, that Man is subject?[35] It does not matter. The discourses of Man divide themselves amid these debates in a smothering confusion, all of them united in a quest for archy-ical human mastery of all the world. Life is lived as either a struggle to become a self-sufficient individual free of social entanglements, perhaps as a capitalist property owner or citizen of the liberal state, or else it is lived in the struggle against individualism and for the community, a socialized consciousness, as a goal governing all achievement, usefulness, and possibility of meaning. Individualism and collectivism are doubles of each other, complementing and constituting each other in their differences, united in the knowledge of Man as archy-ical subject, the underlying reality of all that is. The debate between the two is but a debate of the time, containing no possibility of a decisive encounter with the realities and dangers of the modern world.[36]

Within the dialogue of Man a strange and revealing irony occurs. The more Man conquers the world and becomes its subject, making himself ever more the center around which the thing orbits, the more subjective and impetuously arbitrary his observation of it becomes, for in the thing Man encounters only himself, the one who, as its archytect, imposed upon it its order and rank. His care to draw a line of difference between himself and the thing, so that he can be its master, comes to nothing, and comes to it all the more certainly the more efficient and complete his conquest of it. The observation of the world becomes an observation of Man, science becomes anthropology, and humanism becomes the doctrine of the age.[37]

Only possible in the world as picture, humanism is a moral-aesthetic anthropology, according to the Thinker—an anthropology not in the sense of an investigation of Man by a natural science but in the sense of an archaeology of Man that explains and values all that is according to Man's subjectivity.[38] Since the eighteenth century, anthropology has increasingly set its mark on discourse, and the proof of this is that whatever is, is interpreted as a worldview, an exclusively human Weltanschauung. When the world becomes a picture for Man, a subjective production, the position of Man as viewer, constituted and limited by his historical situation, becomes a worldview, one subjective view among many possible. It goes like this: You have your opinion and I have mine, and any disagreement we have between us is not that important because each position is but personal preference, something that we choose to have, not universal truth. We are each the masters of our own beliefs, and since we are polite people, we respect each other's territory.

This insistence on individual subjectivity is, no doubt, the impetus behind the liberal state's tolerance for different religions and its protection of speech and assembly.[39] Such tolerance was impossible for the Greeks, who dismissed "other worldviews" as barbaric, the noisy babble of outsiders; it was equally impossible for the Christian Middle Ages, which labeled them sin and conquered them with the salvation of true teachings. But the liberal state's tolerance is a limited tolerance; seeking to expand the reign of subjectivity by drawing a sharp line between subjectivity and objectivity, it is unable to accept the truth of any opinion that does not consider itself subjective. It cannot because such beliefs are chosen, a matter for the will, and not for external verification.

Liberal tolerance was a result of a realization that the religious persecution that happened in such events as the English Civil War and the St. Bartholmew's Day Massacre was not about truth, not the kind of truth that worldly science could establish, but about subjective preference. The participants may have been convinced of the truth of their views, but the age was convinced only that they were subjective and that by treating them as subjective, people would become more tolerant. In the age of the world as picture, only that which secures, organizes, and totalizes Man as the guide and underlying reality of all that is, is allowed to expand to its full extent. The world as view contains all safely within the iron cage of its tolerance, making the world secure for Man's control.

CHAPTER FOUR

Technoarchy

The USA slowly lost its mandate
in the middle and later twentieth century
it never gave the mountains and rivers,
trees and animals,
a vote.

—Gary Snyder, *Turtle Island*

ODERN TECHNOLOGY is a revealing of truth, a way in which the world worlds, the thing is brought forth and assumes its identity. Unlike the truth of the early Greeks, and almost certainly the truth of the Minoans, modern truth does not let the thing be itself, shadow and all, but instead imprisons it, demanding that it become Man's utility. The old windmill supplies energy for humans, but it need not take from the wind anything that the wind, as physis, does not freely give. Its sails may spin gently in the summer's breeze, becoming a blurred whirl in the winter blizzard, until its tail sail turns it to meet the spring winds that come heavy with the rain for the growth of the towering tree, the blooming flower, and the grass that feeds the deer. It need not disturb the north winter wind that brings a numbing frost, and it can let the summer afternoon shower go its way after wetting the wildflowers of the field that grow around it. At peace with the seasons, it need not steal energy from the air currents and make them into something other than they are in order to be.

In contrast, the strip mine must imprison the earth, make it totally subject to the human command, before it puts out coal for Man. Locked up in the prison of a purely human archy, the earth becomes a coal mine for Man, the soil a mineral deposit for Man, the sky a place for Man to dispose of the wastes made by burning coal. Indeed, all the earth becomes a coal mine now because it is necessary for it to supply the energy to spin the turbines that supply the electricity to power the factories that supply the products that the consumer consumes that must be consumed in order for the worker to have work and for the owner to make money to pay off her debts to keep the economy healthy so that the state will have a sufficient tax base to build the war machines that are necessary to protect the coal mine. Nothing is left as it is or is unchallenged by the mastery of Man, the modern archy, who, having represented everything to himself as his object, has the truth of all that is at his disposal. Originating in Man's representations, subjected to Man's command, everything becomes available for Man's utility.[1]

The prairies of the American West that the Indians formerly hunted upon and dug for roots, the fertile fields that the Indians of California cultivated and set in order, now appear differently, are different, than they were before the white man came and imprisoned them in his technological utopia.[2] According to an old holy Wintu woman:

> The White People never cared for land or deer or bear. When we Indians kill meat, we eat it all up. When we dig roots, we make little holes. When we built houses, we make little holes. When

> we burn grass for grasshoppers, we don't ruin things. We shake down acorns and pinenuts. We don't chop down the trees. We only use dead wood. But the White people plow up the ground, pull down the trees, kill everything. The tree says, "Don't. I am sore. Don't hurt me." But they chop it down and cut it up. The spirit of the land hates them. They blast out trees and stir it up to its depths. They saw up the trees. That hurts them. The Indians never hurt anything, but the White people destroy all. . . . How can the spirit of the earth like the White man? . . . Everywhere the White man has touched it, it is sore.[3]

She is right: our agribusiness sets upon the soil and, as if it were nothing more than the use made of it, demands that it produce fiber, protein, vitamins, and calories.[4] Herbicides, pesticides, and fertilizers disrupt natural cycles, leaving the soil a poisonous chemical vat. The air is set upon to yield nitrogen and carbon dioxide; dammed and diverted, water is set upon to yield its power of growth; and the soil is set upon to yield its produce, then transformed into saline seep or eroded away.[5]

The truth of things as representations projected by Man and made available as calculative utility is limited, constrained, and twisted into unanticipated forms by the regime of efficiency which Man subjects himself to. The coal that has to be hauled out of a strip mine is not sent for the fun of it, at the whim of its owners; it is supplied because it is useful, needed by some industrial process. It is stockpiled; that is, it is at the call of the system of producing things as a totality. What it is as a thing is contained and ordered within a system of efficiency and rationality that greatly exceeds it, one that is driven by economic and technical demands, social and political imperatives, that cover the whole earth.[6] Because of the dangers of inefficiency to the global system of control, everything—every plant, mineral, animal, energy source, and human being—must be drawn into the systems that reason has built and rendered productive. Nothing can be left by itself; the security, rationality, and well-being of the whole system of control demands that it be brought into its ordering.[7]

The archy that rules over the destiny of things throughout

modern technology has a demanding, arrogant, and conquering character, challenging things to come forth and be as they must be for Man as technical master of the earth to continue to his destiny. The energies concealed in nature are challenged forth as Man's object, representation, and utility. They are transformed into the forms most useful for him, stored when and where it is necessary for him, then distributed to serve his end uses, which because they are many, interdependent, and complex, are switched about ever anew. Unlocking, transforming, storing, distributing, and switching about according to the imperatives of Man's reason are ways of revealing the truth of Man as father, origin, and command. As the modern archy, the one whose representation presents everything that is, whose command everything submits to, Man is that for which everything is organized, done, and accomplished.[8] It is his choice. And yet what is there to be organized, done, or accomplished? Nothing except Man, the patriarch's, possibility of command. Since there is no other origin or command to bring things forth except Man, there is nothing to govern the bringing forth of things except the possibility of Man's patriarchy. Nothing is for anything except command, more power, more control, more exploitation. Without anything to use all this power for, everything thus becomes a means for something else. The revealing never comes to an end, a goal, or a purpose. Man may, at last, be subject, patriarch of all the world, but his subjectivity is empty of content, of anything to do.

Is the end of the Rhine locked into supplying power for production in factories for the consumption of heat and light in the home, or is it there for the profits of the electrical industry, or the steel and chemical industry? Then again, might it not end in the security of the German state?[9] All these ends, these things that it all might be done for, fade into each other and are revealed as intermediate, dependent, and lost in an aimless confusion that must be mastered. Nevertheless, it all must be maintained, kept going. In order for Man to keep his position as master over things, he must regulate and secure things as they move about their interlocking paths, making sure they follow the most efficient course, the way most likely to preserve everything's utility for everything else. So much is this so, that reformulating, regulating, and securing become the chief characteristics of reason's way of revealing.

In the unconcealing of modern technology, all things are revealed in the world as reserved, ordered to stand by and submit to further ordering by Man and his reason. To be there for something else, whatever it is. To become whatever is needed. As the objects of Man's will, they no longer present themselves to us as things, gathered together by a care that is near at hand and responsive to the earth, but rather they come forward as constituent parts of a technical system that orders them over vast distances and compels them to revise themselves continually according to Man's patriarchy and to fit within the shifting contexts of whatever command is commanded.

Hegel defined the machine as autonomous tool, immediately ready for the use that can be made of it, something that stood by itself and was for something, and the Thinker agrees that when this definition is applied to the tools of the craftsman, it is appropriate because the craftsman, responding to the calling of the earth, uses her tools to respond to the thing before her, quite independent of any other consideration.[10] The tool is for something specific, nameable. But this definition does not reflect the way machines are now. Revealed as the Reserved, the machine is not autonomous, is not for anything in particular, any good; it does not respond to its place on earth, only to the placeless command. Shifting, pointless, and lost amid its revealing, it belongs to a technical system that itself is without center. In this, the modern machine or tool is sharply at odds with itself. It is an impossible contradiction that, despite its impossibility, nevertheless is.

A tool is something that is used *for* something else, a means for bringing a thing forth. It belongs to a situation which determines what the tool is—how it is made, how it will be used, and when it is worn out, how it will be disposed of. Because tools are a means for bringing something forth, they cannot be visualized or known apart from the world that they help bring forth, the place they happen at. An interpretation of their use, their purpose, and their meaning must be available for a tool to be a tool. By themselves they are nothing. A hoe is for tilling the earth and making it possible to bear fruit. Depending on the world that it is used in, it can reveal the earth's power to be fruitful or it can reveal the power of Man to impose his will upon it. A tool always refers to a world, a god as the Thinker would describe it in his more poetic works, which provides an interpretation of its use and gives meaning to its purpose.

A tool is *for something*, yet it is the fundamental character of our age, when everything has become a tool for Man, that our tools are *for nothing*. The thing that the modern tool brings forth, the purpose that gives the tool its meaning as tool, is only another tool, another instrument for Man's willing. The revealing never comes to an end, never reveals a god that would make the tool a *tool* by making it into a *means* for something. This is why nihilism, the purposeless and aimless willing of everything as a means for Man, is the ultimate truth of our age, as also is Man's patriarchy. For us in our use of things, there is no god providing an interpretation of their use, nothing to tell us what a thing is for. Nothing is revealed by all our technology; truth does not happen.

It is Man, the tool user, through whom the challenging of the thing takes place, he through whom the thing is revealed as the Reserved, and he who accomplishes the thoughtless, purposeless, and nihilistic destruction of the earth. But despite his arrogant illusions to the contrary, Man does not control unconcealing itself, he does not decide, will, or cause the thing to come forth as the Reserved.[11] On the contrary—and this is most depressing—the Reserved precedes and possesses Man, granting him its truth as will amid the modern world's worlding. If the world has come to us and been interpreted as the lighting of Ideas since the time of Plato, it is not Plato the writer who is the cause of this—he only responded to what presented itself to him. If the world now is, in its entirety, a tool for our use, it is not that way because we have willed it that way, but because that way is the way that has possessed us.

Only because Man is already possessed by the ordering challenging that conquers the world as the Reserved does he dwell amid it as such. And if Man is possessed by a world that is present as the Reserved, he himself becomes more profoundly and more terribly the Reserved, the object of his willing.[12] As John Kenneth Galbraith has written:

> Our wants will be managed in accordance "with the needs of the industrial system"; the policies of the state will be subject to similar influence; education will be adopted to industrial need; the disciplines required by the industrial system will be the conventional morality of the community. All other goals will be made to seem precious, unimportant or antisocial.[13]

Fulfilling this need for control over humanity, we see human labor replaced by machines, because it is more gainful, and the unemployed treated little better than worn-out machines, a useless clutter only grudgingly left any space at all.

The modern farmer may grow some of the same crops, work the same fields, and live in the same house as her grandmother, but unlike her grandmother, who grew a diversity of crops primarily for herself and her family and only secondarily for profit, the modern farmer is commanded by the aimless imperatives of profit making in agribusiness. As a businessman or a worker, she is subjected both to the market that organizes and fabricates the demand for food and to the market that supplies her with the technology of production that she uses. The food that she grows is grown not for the health of her family or of those who she trades with but for the cash necessary to pay off her debts and her interest on them, and to buy the equipment, fertilizer, chemicals, and fuel necessary for a cash crop. The cash crop bought from the farmers is, again, sold not for the health of the person who eats it but for the profits of the food-processing, transportation, advertising, and retail industries. Because health is not the concern that gathers the crop up and sets it onto the table, many dangerous chemicals and additives are present in it to secure its profitability.

So much has the health and nutritive concern of food making disappeared from its production that the cost of the wrapper for a loaf of bread is greater than the cost of the wheat in it. But this is not surprising, because the function of the wrapper as advertisement and preserver is to ensure that its consumption is available on demand. The truth of the consumer as Reserved for the food industry is again demonstrated by the monstrous actions of the Nestlé Foods Corporation in the so-called underdeveloped world.

A dramatic reduction of demand for baby formula, one of Nestlé's products, followed the end of the baby boom in the United States. To keep its factories going and its profits high, Nestlé began an advertising campaign throughout the underdeveloped world. Bottle-feeding was sold as a "modern" technique, much more sophisticated than breast-feeding. However, because of conditions in the underdeveloped world—low per capita income, poor sanitary conditions, ignorance about modern technology—bottle-feeding turned out to be entirely inappropriate, resulting in the death of tens of thousands of babies wherever it was used. Mothers could not afford the amount of formula necessary, so they diluted it with water, often unsanitary water, thinking that would be OK because the formula's powers, as they were presented in the advertisements, were so great. The result was malnutrition, disease, and death.[14]

Both the farmer and the consumer are revealed as the Reserved by the food industry—the consumer because her consumption is a thing to be manipulated and controlled by advertising technology, the farmer because her craft is measured by its usefulness to the food industry. "Inefficient" farmers go broke and become surplus farmers. The ugly truth of the farmer as Reserved by the food industry is best revealed by President Reagan's joke that we should keep the wheat and export the farmers.[15] From time to time, though, the farmer's work is dignified and surrounded with respect when her cash crops for export become a weapon, a tool to punish and control Communist and Third World countries that resist the will of our foreign policy managers.

Yet Man is not gathered into the Reserved as the energies of nature are because he is not passive before its onslaught. In subjecting the world to his will and making it into the

Reserved, Man subjects himself to his reason and his destiny, driving technology forward. But the world itself, within which the ordering challenge unfolds, is not and never can be a human handiwork. The world worlds whenever humans open their eyes, unlock their ears, and attend to what presents itself to them.[16] In their work and their living, their meditations and their entreating, humans reveal the world, bringing its things forth from unconcealment in the earth. When humans reveal the thing, they are merely responding to the calling of the world, the breaking forward of the earth, even when they contradict it.

When Man, investigating and observing, ensnares nature within a web of his own pictures, he is already claimed by a way of revealing that requires him to approach nature as an object of research. And nothing will stop this way of revealing, even if, contradicting itself, the thing disappears into the nihilistic thinglessness of the Reserved, becoming no-thing in the endless web of reason's revealing.

Ordering and revealing the thing as the Reserved, modern technology is thus not anything Man does. He is gathered into it, captured by the world's worlding. The gathering of the thing, the truth of the world specific to the modern age of technology, the Thinker calls *Ge-stell*, or "Enframing," as it is commonly translated.[17] Technoarchy, as I shall read, translate, and think it in order to draw it near to my own life, is the originary claim, the command and interpretation of the thing that governs the world, revealing it as the Reserved, a mere tool for Man's aimless patriarchy. The commanding origin enframes, in that its interpreting assembles and orders things into being, trapping everything that it brings forth in a framework or system, ordering it for a use that is always restructured and revealed anew as Man's will.[18]

Nihilistic throughout its whole extent, Technoarchy is the way of revealing that is the truth of modern technology, but it itself is nothing technological, since it, in its originary truth, is there before it is, and it is in no way anything mechanical, calculative, or procedural. Making the calculating machine calculating and rational procedure rational, it is what precedes

Man's will and reveals the world as the Reserved, the truth producing the thing and presenting it as Man's utility.[19] A truth preceding Man, it is therefore not a human activity or a mere means within such activity but a way of Being governing how things are brought forth, a truth that knows everything as a coherence of forces available for exacting calculation and measurement. The truth of technology is not represented by gadgets, exotic tools, and technical accomplishments such as the computer, the space shuttle, or the nuclear reactor; rather, it is present as a ruling command that sets Man up as its origin, the patriarch of the world.

Contrary to popular opinion, far more radical revolutions in technology separate Parmenides from Aristotle, and the sixteenth century from the eighteenth century, than anything that has occurred in the last century. Between Parmenides and Aristotle and the sixteenth and eighteenth centuries, entirely different ways of Being broke forth from the earth and came into their truth, separating the past from the present with a radical gap of incommensurability.[20] Amid all the supposed technical revolutions of the modern age, there is in fact complete continuity. All the technical accomplishments of our time have occurred within a way of Being that has in its truth remained the same.

Our way of worlding was first revealed in the rise of modern physics as an exact science. As we saw, modern science's view of the world as picture, its separation of the world into subject and object, pursues and entraps the earth as a calculable coherence of forces. It is experimental not because it is first and foremost empirical but because it projects nature as something mathematically calculable and measurable in advance. The possibilities of the experiment—that is, the variables formalized within its structure—are ordered and structured in advance so that knowledge can be gained of how nature reveals itself when set up in this way.

But modern mathematical physics existed almost two centuries before the revolutions of modern technology started occurring. "How, then," the Thinker may well ask, "could it have already been set upon by modern technology and placed in its service?"[21] Although modern technology advances with the aid of the sciences, it does so only because the way has been prepared for it by the modern physical theory of nature. And this is precisely what the truth of modern technology does; it reveals things, pictures them or objectifies them, in a way that makes it possible for the sciences to know them. Although modern physical science began in the seventeenth century, and machine-power technology developed only in the second half of the eighteenth century, the truth of technology was revealing and holding sway from the very start of the modern age.

The calling that gathers Man into Technoarchy is his destiny and his doom—his destiny because from the beginning of our world, he is sent on his way to it; his doom because the way to it obliterates all that he is in his truth. But destiny is never a fate that merely compels, for it is only in meeting its destiny, following the ways of the world worlding, that humanity finds its truth and its freedom.

As I said before, the truth of freedom is not connected with the will, collective or individual, or even in the causality of human willing. People do not come to their freedom by means of arbitrary choice, neither do they come to it in the constraint of law, nor do they come to it by means of a demystified, fully rational, and reflective consciousness; they come to it by sparing the world, preserving the mystery of the earth, and accepting their life.[22] People are free when they let truth happen.[23] Anarchy describes this way of being. Seeking neither to command nor to trace its origin back to a universal and eternal principle, anarchy is a freeing presencing that conceals in a way that opens the thing up, letting the mysteries of the earth be what they are and the visible openness of the sky be what it is.

As Technoarchy, modern technology, too, is a way of revealing things, of coming to truth. It is our destiny, even if it denies itself its own truth. But, according to the Thinker, letting the world world does not mean that we must blindly and thoughtlessly push on with technology, nor does it mean that we must reject it, rebelling helplessly against it and cursing it as an endless evil. Either action is only a continuation of the will's willing, a judgment that Technoarchy is in full accordance with. However, when we open ourselves up to the truth of technology, knowing it as a way of revealing things, we can free ourselves of its relentless logic and meet the things it reveals as they are—not as the objects of our will, but as the earth breaking forth. It is a lot like a Zen meditation. By accepting the world as it is, we change it. By trying to change it, it stays the same—especially in this age, caught, as it is, in the tempest of Man's willing. Nothing could be more subversive to it than to stop trying to will things as they should be. By letting the world world, it will world in an entirely different way than it does.

If we stop our willing and let the world world, we find ourselves placed in a world open to Technoarchy, closed to anarchy, and ourselves endangered by our own destiny. This different perspective changes things because then we know the destiny of Technoarchy is a danger and a possible doom. The Thinker tells us that while Man may well destroy himself and his planet with the awesome power of his machinery of war and production, it is only because he is already dead—dead because his truth is concealed from him and because he pictures himself as only his will. By letting this death be, we can see how, seeking to be masters of the earth, we have failed our calling to be the guardians of the world, the friends of the earth. And we can redeem ourselves.

We will need to do it ourselves because God is not there anymore. When earth's things come to presence in the light of a cause-effect coherence, even God loses all that is exalted and holy, and the mystery of his distance disappears. In the light of patriarchal causality, God becomes merely the first cause, the mechanic of creation. Even in theology, God becomes the God of experimental science, his mystery concealed by the causality of making.[24] In America nowhere is this more evident than in the recurring battle creationists have with evolutionists. Evolution, it is asserted by our devout critics of modern science, did not cause all the plants and animals and especially Man to come into being—God did. Indebted to the science of causality, yet rejecting its scientific truth, such a God as the creationists have can be maintained only with the most profound hypocrisy, and such a God is dead. Why else does Jerry Falwell demand frequent loyalty oaths from the teachers at his Bible college?[25]

When the world's destiny is Technoarchy, it is in supreme danger. The danger comes upon Man in two ways. When things are present no longer as the earth's mystery but exclusively as the Reserved, and when it becomes Man's destiny to be the master of the Reserved through his reason, he comes to the brink of a catastrophe, a precipitous fall—he himself becomes Reserved, the object of his own willing, a tool for his own utility.[26] Meanwhile, as the one threatened, Man exalts himself as the patriarch of the earth, the one whose constructs and representations constitute things as a coherence of forces. This humanist illusion deludes Man into believing that in everything he encounters he meets only himself. But in truth, Man nowhere encounters only himself, his truth. So totally is Man locked into Technoarchy that he does not understand it as a calling or understand that he is possessed by it and that he is the one spoken to. He fails in every way to dwell as a mortal letting the world world, the earth be. He thus can never encounter only himself, because what he is, the guardian of the world, is concealed from him by his way of being.

Technoarchy not only denies Man his truth, throwing into oblivion his relationship to himself and to everything that is, it subjects Man himself to its ordering, driving out every other possibility of revealing by its discipline. Through its compelling demand for order and reason and its subsequent fear that makes it throw everything that is not orderly and reasonable into the shadow, Technoarchy conceals and represses any sense of a revealing that lets things rise up from the earth and come forth of themselves. The shadow must not be allowed to be. Ordering, regulating, and securing things as the Reserved, Technoarchy denies things their own character, their earth-born identity. Fearing its shadow—what it has dismissed, repressed, and made other—and not letting it be, Technoarchy not only denies all things and all humanity their truth, it conceals revealing itself, the way truth happens.[27]

Ordering the world about as its utility, Technoarchy locks up all truth of things into its ordering, denying them any possibility of becoming present as a gift of the earth or of letting them hold sway over humanity as such. The destiny that sends us to Technoarchy is consequently the extreme danger. But it is not technology itself that is dangerous, the machinery of destruction and production that have given Man so much power over the earth. They are not the danger of Technoarchy; it is its truth of things, the truth that keeps humanity from its calling to be Guardian of the Earth, that risks making our destiny into a doom.[28] Machines of limitless destruction and production are nothing and hold no sway unless they are ordered into use. Ordered as the Reserved, compelled by the necessities of reason, haunted by their shadow, these machines are useful in a world that uses them.

They do not have to be that way. They could be quite other than what they are now in a world that lets truth happen.

CHAPTER FIVE

The Flight *of the* Gods

Hopes and Dreams
lie shattered on the floor
Broken bits of glass that
tear the flesh
torture the soul

For a while despair . . .
Then, timeless and dark,
the endless night rolls in
And Life becomes
A whisper echoing
in ears that do not hear
A dream that is no more

—the author

IF THE WORLD ever becomes all cold and dark, if death ever becomes a tidal wave that swirls over all the earth, it will not be our greatest tragedy, only our last. Before that, preceding it, making it possible, was a destiny that called us to our fate, a destiny that possessed us before we were born and now locks us ever more tightly in its grip with every attempt we make to escape it by willing the world otherwise. We cannot overcome the destiny of our age with yet another triumph of our technology, yet another assertion of our mastery, because any attempt to escape our fate through experimental research, calculating reason, and technological triumph will only serve to advance its cause and assure its triumph.[1]

Herein lies the tragedy of our age. For it truly is not our lack of will, our inability, or our negligence in mastering the things we have not yet controlled but our method of solving all our problems, our way of technical mastery itself, that shall lead us to our tragedy—if we come to it.[2] Our very strength, the self of power and solution that our scientific and technical discourse gives us, is our weakness, the danger threatening us. It is this self, knowing that it is not always the master and interior cause of its actions, that feels that it must know the ways it is dispossessed of sovereignty and then, by becoming master of the causes formerly external to itself, repossess its patriarchy and assert its power.[3] Suspicious that there is always a cause or power external to itself, fearful of any shadow it has not banished, obsessed with being the master of itself, the modern patriarch is possessed by a grim and all-consuming will to truth and mastery. Before it all actions, thoughts, inclinations, and ugly secrets must confess their difficult truth; before it all things must pass in review and be interrogated with a cold ruthless zeal to see if their reality and truth is not something different from what Man has mastered, another shadow not yet banished.[4] All things are held in reserve for reason as something other than itself, some concealed truth or being that must not escape the scientific sovereign.

To our age of power through technology, of mastery through reason, were born three great fathers of suspicion—Freud, Marx, and Nietzsche.[5] Though they are different from each other in their own way (and Nietzsche much more ambiguously than the other two), they are united in a metaphysic, a technology of truth and power, that makes Man the truth of all beings, the fathering archytect of all the world. Seeking to preserve the legitimacy of what Man fathers, the three great fathers of suspicion share this in common: a deep and abiding suspicion of appearances and accepted interpretations of them and a certainty that their truth lies in translating appearances into the language of human desire, in understanding them as a human production, or in conceiving them as the all-too-human deposits of the will to power.

Before each of these fathers of suspicion Man finds himself dispossessed by a shadow he let slip away from his self. Finding his delusions intolerable, he is called on to overcome himself, to grab by the neck the moonless night of all of his lies to himself and to make them into his own will, to make the fatherhood of Man legitimate, subordinate to its true origins. Where the id was, the ego shall be; where there was the mystifying veil of class exploitation, there shall be the worker's paradise of Communism; where there was the rancorous slave, there shall be the overman. The moment of affirmation for all these fathers of suspicion is the moment where Man returns to himself as the abiding power, when the shadow haunting him is removed and he wills only himself, the archytect of the world—the moment when the future classless society and its transparent values become actual, the moment when the ego masters the id and dissolves its pathology, the moment when the revaluation of all values is understood as the positing of value.[6]

(But wait, before I go further, I must admit I am reducing Nietzsche, the most worthy of these three fathers and probably the best poet of our time, to something he is not entirely. If Nietzsche is *the* theorist of the will to power, the one least afraid of openly affirming it, he is also the thinker most able to understand its limits, dangers, and shadows. Nietzsche knows there are shadows in his thought, things left unsaid, things that dare not be said. In fact, unless we miss them, he often points them out. Dots wandering off . . . Zarathustra knows that transparency, pure will willing only itself, is not possible, and he has the courage to face it. Self overcoming cannot end, and we can never become our own fathers. A true poet attending to the earth within, Nietzsche is probably the most courageous thinker to have ever lived, and yet there is still a fear, a shadow that lingers . . .)

What is the fruit of all this fatherly suspicion, this fear that seeks out all the shadows of illegitimacy and wills them away, this attempt to assert control over the revealing of things and make it only its own? It is, as Nietzsche himself first pointed out, nihilism, the radical repudiation of value and meaning, the dissolution of the world Man's reason has built. For Marx, Freud, and Nietzsche, moral truth is approached, not as something that is a response to the need of dwelling or the truth of the stay of mortals upon the earth, but as an object of knowledge to be researched and analyzed, whose genealogy must be chronicled and synchronic structure mapped, whose causes must be calculated and transformations submitted to scientific interrogation and understanding. Knowing everything as nothing but a re-presentation of Man, an archytecture originating in the will, the necessity of legitimate fatherhood reduces everything—all morals, traditions, and institutions—to the will willing itself, to nothing but Man himself.[7] Referring only to itself, without direction or aim, all that this archytecture can reveal is more of itself, more technology, more power, more mastery—more of Man.

Having accomplished the possibility of control, fatherly mastery undermines the value of the things it has conquered. It engages itself in a ceaseless struggle to become only itself, mastery transparent and pure. And yet what institution or value can survive this accomplishment? Once everything becomes Man's value, it can be revalued, reassessed at whim, for whatever reason Man wills.[8] And because the value of everything can be reassessed at will, according to whatever Man at the moment wills, it loses whatever value it could have, becoming totally valueless. It is these three fathers of suspicion, speaking the only truth available to our time and setting themselves up as the sovereigns before whom all values must pass in review and render up an account of themselves, who have spoken the word that killed God the Father for us, vanquishing all the purpose, value, and meaning that he gave.[9]

Just as well too; the patriarchal god they killed was a god that was judgmental, condemning, hostile to the earth and all life, authoritarian, and an enemy to its own shadow. A bad father. Unlike the goddess that preceded him, this god made life joyless, fearful, hierarchical, and exclusionary. Where the goddess accepted and affirmed the shadow as part of herself, this god refused to, projecting it on his enemies, and then making them pay for "their" sins. In their suspicion, their will to truth, these god killers all revealed this god as something other than himself, helping to break the world free of himself and his gloomy reign. But yet, in other ways, they remained deeply indebted to him, perpetuating his ways. I call them the fathers of suspicion because there is something, like the god they killed, very patriarchal about the world they condemn and the world they seek.

Before the age of Technoarchy, before the time when Man became master of all the earth and beyond, God was our Father, the absolute positer of value and meaning, the creator of a metaphysical whole, the seminal origin of a beyond infinitely superior to ourselves. Insisting on his fatherhood, demanding its acknowledgment in humility, belief, and obedience, God possessed our being with his fatherhood, his purposes, and his ends. We were his children, and our life, joys, needs, and connections with each other and the earth could only be in a way that reflected our love for him. Anything else was sin, forbidden. The metaphysical realm God fathered was not anything that rose up from within, built of all-too-human needs situated in our place on earth, but an external whole that was its own necessity, a beyond by which we measured ourselves, judged the world, valued things, and lived our lives by. Living in fear of it, we were permitted no participation in its being. Father knew best.

The realm of the metaphysical was like the light of Plato's sun shining down on us, lighting up our practices and actions with the Father's meaning and purpose. Afraid of God's judgment and punishment, it was our guide and our hope. But then, technical Man, swallowing up the sea, wiping away the horizon, unhinging the earth from the sun, put himself in the Father's place of authority.[10] Now, as the technician of value, the judge of all conventions and practices, the sovereign of the new scientific truth, he rose up, as subject, and transformed everything, including the metaphysical realm, into a scientific object to be prodded and turned over, analyzed and dissected[11]—to be fathered by him. Morals became conventions, beliefs became values, faith became a delusion, and all were held in reserve for the social technologist, the propagandist, the advertiser, the revolutionary—to work upon, to transform, and to make more rational, more useful.

Casting God down from his throne of authority, we became the technicians of all value positing, and the value we, as masters, have of things becomes the truth we impose on them. We became our own fathers. In our age the sun no longer emits light of itself, free of our meddling, but is instead a dull and dimmed moon, getting what light it gets from us, the masters of the earth. Once autonomous, the metaphysical realm is now nothing but a human point of view, posited by the will to power, a production contingent on the mode of production, a fantasy made necessary by repressed desire. The scientific sovereign of our age seeks to surpass, to overcome, Man's finitude up to now, making itself over into the father of all that is, drawing itself and everything else up into its own willing, a pure will willing only itself.[12] All that is, is as that which originates in the will of Man. All the truths or gods that formerly conditioned and limited the truth of Man, providing an interpretation to his life, his actions, and the things in his life, have fled the earth. They, and especially the Christian patriarchal god, are dead.[13] As the suspicious father of everything, Man has killed them, plundered the temple, and desecrated all that is holy.[14] Far removed from the earth, lost in empty patriarchy, the world has become lifeless, dead, a circle without a center.

Too true, there is Christian faith here and there. And sometimes it rises up with fanatical strength, but the love and, more often, the fear that sustain it, is not the effective, determining truth of our time.[15] No modern state listens to the will of God, and in America, the prototype of the modern state, anything associated with God is rigorously excluded from the actions of the state. No multinational corporation responds to anything except profit and power. In our work, our making, our living, the Father's holy word, what our being was for, is absent. With its disappearance, all the things of the world lose the Father's judgment—the trees, the flowers, the birds, the animals, the sins of man and woman, the nature of their works, and everything else,[16] including all morals, traditions, and institutions.[17] No longer is the God, the divine archytect, the underlying truth of things. This is what Nietzsche means when he says that God is dead. It is not a claim about God's existence or nonexistence but an assertion that God is irrelevant to this age of Man, no longer near to us in our life.

The history of the West is the history of the progressive concealment of Being, the silencing of the earth. Where once the name for Being was *physis*, a blooming forth of itself, after Socrates, after the Romans had mistranslated Being as *natura*, after Christianity made knowledge of Being into a dogma and a metaphysic, Being appeared instead as archytecture of action and reaction, more and more a chain of causes and effects available for reason. Along with Platonism, Christianity is initially responsible for withdrawing Being from the earth and its anarchy, where it was revealed to the early Greeks, and removing it to a metaphysical beyond.[18] The truth of things moves beyond *(meta)* the earth *(physis)*, becoming metaphysical.[19] The earth is closed off, and Man is no longer confronted with its mystery. Cut off from the earth as the revelation of Being, lost in its archytecture, Christianity turns to scripture, which silences all questioning before its dogmatism, and to metaphysics, which elevates categorical understanding as the way of knowing Being.

Although it has become the history of world, Christianity need not have turned to metaphysics and dogmatism. According to the Thinker, the apostles John and Paul, Augustine, Aquinas, and Luther still were able to experience Being as the abyss of God, to feel the holy rising up through them, within themselves. As soon as they attempted to explain their experience, however, they changed it from an inner experience of connection to an outer experience of domination, and they became entangled in metaphysics and dogmatism.[20] Some Christian mystics, like Meister Eckhart, managed to avoid the archytecture of dogmatism and metaphysics by using contradiction and tautology, but they were unable to find any listeners. Instead, they became saints, vehicles for the church to expand a new dogmatism. Perhaps in our day Wendell Berry is a Christian who has not succumbed to the traps of dogmatism, scripture, and metaphysics. Perhaps he is able to attend to the erupting earth within himself without judgment or fear, free of the Father's archytecture. But he is almost unique in that.[21]

Although Christianity has occasional flashes where it has access to Being as an inner revelatory enigma, such flashes are quickly overwhelmed by the dogma of metaphysically interpreted scripture. As a result, God is not a way of Being, a world whose worlding calls on Man to encounter the mystery of the earth, but a particular being who has created both Man and nature and who has revealed himself in, and as, Christ. As the father of all beings, the archytect of the world, God is understood by dogmatic Christianity as merely the highest and most real being, the first cause, the seminal creator.

Being itself is ever more completely interpreted as archytecture, as causality and fatherhood, retreating ever further from the human dwelling place. As a result of the earth's being entrapped in metaphysics, dogmatism, and archytecture, the fundamental way of Man upon the earth is as a faithful believer. Cut off from the inner experience of the earth by metaphysics and dogma, lost in the Father's archytecture, the only thing left to humanity is faith, the abandonment of all questioning, all thinking.[22] Having no access to God except scripture and the certainty of faith, the believer's God must be incomprehensible, beyond all human understanding, because only faith can make the impossible leap between the earth and the beyond.[23] It is a distant god, this patriarchal god, far removed from the dwelling place of humanity, the daily cares and experiences of mortals—a god that is indifferent to limits, judgmental of fault, insistent on commitment, impossible to know, jealous of alternatives. It is within this archytecture of the distant God that Technoarchy and the danger of nihilism approach.

Technoarchy appears first as the doubting of the legitimacy of all traditional authority, and especially the authority of scriptural revelation. Though doubting everything, Man nevertheless turns not to the earth but to himself as the certainty of everything, leaving the basic archytecture unchanged. He becomes the legitimate father of all beings by becoming their true origin and master, claiming everything, body and soul, as his own to do with as he wills. Even when God has died, Technoarchy remains fundamentally Christian in its way of Being: the certainty of faith becomes the certainty of doubt, divine creation becomes modern technology, and throwing God down from his place of authority, Man himself becomes the highest being, the prime mover, the final end of the world.

Even when God is dead, we remain good Christians, faithful sons and daughters. We will not escape this archytecture by being blind to it, dismissing it as the final delusion of a history that for too long thought that God mattered. Despite ourselves, lurking always at the corners of our thought, our dead God will remain as a shadowy guest, all the more pervasive and influential the less he is thought.[24]

According to the Thinker, we do poor service to thinkers when we but repeat their thought, interpret it, or extrapolate it correctly. We do thinkers honor only when we think their thought, listen to the earth breaking forward in it, and draw it near to our own life. And this means that we think about what is near in their thought, that we situate it in our lives and direct ourselves to the unthought in their thought, thinking and perhaps living their thoughts more deeply than they. That is how we free ourselves of the archytecture of metaphysics. As he does this for Nietzsche, the Thinker does not limit himself to a correct interpretation of his thought but situates it in the world that worlded Nietzsche and his thought.[25]

He finds that there is yet more shadows in the depths of Nietzsche's nihilism than Nietzsche himself thought, that despite himself, Nietzsche has not overcome nihilism but only served to extend it, rebuilding its archytecture.[26] According to Nietzsche, the death of God makes possible the overman, the one who will revalue all values and make himself into master of the earth. This should not simply mean that Man, in the form of the overman, directly takes the place of God, usurping his position as he dissolves his authority, because this would not be thinking in a holy way about the holy.[27] Mortal and all too human, Man cannot put himself in the place of God because the truth of Man does not reach up to the realm of the holy. Instead, according to the Thinker, something more uncanny happens. Thought metaphysically, as it has been throughout the entire history of Christianity, the place of authority belonging to God is as the cause and preservation of everything that is. Everything belongs to God because he created it. But this place can remain empty, and instead of being occupied, another place, another archytecture, corresponding to the age of the overman, can appear in the sky—a place that is identical with neither the realm belonging to God nor the realm of humanity under his authority.[28]

The place that Man comes to occupy in his time as the overman is unique to itself, bearing no relation to what was. Despite dissolving his authority, the overman does not usurp the place of God; rather, he lives in another world, another time, and another truth governing the way of things. This other archytecture is subjectness. Everything which is, now is for a subject, an object re-presented before an ego cogito. Viewing the world as picture, a re-presentable object, the ego comes to be its own archytect, like a mathematician positing the form and structure of things. Returning to itself from its object, the truth of consciousness is self-consciousness because it re-presents itself as the world to itself. In doing so, it becomes its own father, its own archytect. Everywhere, for every modern thinker, Nietzsche as well as Descartes, the archytecture of whatever is, is as re-presentation, a subject setting itself before itself, becoming its own father, its own cause, origin, command. Worlding, the world is viewed as picture, as re-presentable object. Everything is delivered over to Man's fathering or re-presenting, putting it in the midst of Man's positing, calculating, and mastery.

Since it bears within itself resistance to the father's will by being other than it, the shadow to its light, the earth itself is revealed as the object of an assault, an aggressive challenge that draws everything into its circle. Resisting the claim of

Man's fatherhood, nature becomes an object to be dominated by Man and his technology, and the will to power becomes the truth of all things. As the world comes to be as object and the struggle for dominion over all the earth becomes a frenzy, the age of human subjectness, having revealed the lies underlying all other archys, is driving itself to self-consciousness and to its final archy, the will to power itself.

As we saw, patriarchy is primally experienced as a lack of control over revealing, a fear that it is not in control of things, and so, motivated by its weakness, it seeks control, power, the right to define, limit, and claim. Fearing for the legitimacy of its claim, it builds a succession of archytectures to assure it. At last, driven by honesty to its truth, it comes to rest only on itself, the sheer power to make its claim. The completion of this truth about all things is becoming certain and conscious of itself as master, the archytect of the world.[29]

According to the Thinker, self-consciousness is the necessary instrument of the willing that wills as the will to power, whether it is the therapy-mediated consciousness of Freud, the class consciousness of Marx, or the feminist consciousness of sex as the artificial and Man-made fabrication dominating feminine lives. Objectifying the world and everything that denies it its completion, self-consciousness as a way of Being becomes necessary for economic planning, for rational childrearing, for psychiatric therapy, for a scientific theology, for the liberation of the oppressed. Obstacles must be overcome, traditions modified, irrational values and superstitions disposed of, and contradictions resolved before control is complete and fully rational, fully willed. The quest for self-consciousness finds it necessary to dissect history ceaselessly, to objectify and interpret it as one archytecture or another Man has possession of. Once this correct interpretation of history is made, reduced to the entirely human thing that secretly determines it, underlies it, and shapes its destiny, then it can be grabbed hold of and made subject to human will, making Man into the world's archytect.

Despite his critique of the subject and of reason and science, despite the fact that he knows that pure transparency of the will is impossible, the thought of Nietzsche represents the "great noon" of human subjectivity, the time of brightest brightness, when consciousness becomes conscious of itself as the will to power. At last unashamed of itself, willing itself as the will to power, the age of the overman objectifies everything in its world, making it all totally and uniformly secure as something reserved for itself. But as the will to power wills itself, it must also will the history that makes it possible, every moment of it as the Eternal Return of the Same. To will everything as it is, even the most horrible of moments, the most terrible of conditions, is to accept and profoundly affirm the consciousness that reveals the will to power as the truth of everything. At the moment that everything is willed as it is, as it was, as it always will be, the will to power becomes itself—conscious of itself as the will to power, attaining its highest freedom, the freedom to be the master of the earth. And Man becomes father of himself.

At high noon, when the will to power affirms itself and is conscious of itself as the will to power, the world worlds as value, as something posited and affirmed by the will to power. All things become valuable, placed in a monotonously exclusive relation with the purely human will that evaluates and places them in a relation to itself. Valuing becomes the archytecture of the entire world, the underlying truth of everything as the Reserved. As the thinker of the will to power, Nietzsche is the best expression of Technoarchy, however ambiguous his relation to it is as a poet. It would seem that the world could not be more highly esteemed than being as value—and positing the world as value is how Nietzsche overcomes the valuelessness of all values—yet, according to the Thinker, this is not true.[30] Worlding itself as value, the world is only degraded more. Nihilism is not overcome; it is extended to its terrible, uncanny conclusion.

When the world worlds as value, its truth is sealed off from itself, obliterating every way of experiencing the mystery of Being, the erupting power of the earth.[31] But this has long been the destiny and danger of Western thinking.[32] Since Plato, since Christianity, Western thought has been marked by its metaphysical closure, its inward necessity for making the

Being of all things a being, for interpreting the inward truth of a thing as a metaphysic or archytecture of some sort—the one God, the logos, transcendental reason, sexual drives, class dynamics, or whatever—and for cutting itself off from the earth. When finally, with the triumph of technology, the world appears as value and power, rendering valueless all previous metaphysical systems, metaphysics is only fulfilling its destiny, the complete concealment of the earth.

In our age, when technology has made us the master of everything because it is held as the Reserved and is as value for us, the oblivion of Being, the concealment of the earth, the dark night of the world, haunts us unforgivingly. Closing inward on itself around Man, throwing everything else into the shadow, the world worlds nothing, no-thing. Held as the Reserved, everything loses itself in modern patriarchy, an aimless, groundless willing of the will. Nothing permeates our lives, our work, our things, our gods, and beckons at us from our graves. Subjected to control, everything has lost its center, its purpose, the metaphysic that it is for, disappearing in a uniformly and infinitely distant shadow of oblivion. Nothing, no-thing, is near at hand, close to the dwelling place. And thinking in terms of values only draws it nearer to hand, becoming pure nihilism. Nihilism, as the revaluing of all values hitherto, is overcome as an affirmation of valuing, only within its own archytecture. Taking the will to power as the truth of everything, valuing does not let the world world, allow Being to be as Being, or let the earth bring forth its mystery, but instead brings about the consummation of nihilism. Man's patriarchy does not let truth happen.

Not only does this metaphysics of the will to power not think Being as itself and not spare it its identity, it overturns all metaphysics, setting itself up as the refutation of all previous metaphysics. But the reversal of metaphysics, of Platonism and Christianity, the devaluation of the metaphysical and the revaluation of the physical, is still determined, and remains identical with what it overthrows.[33] It remains metaphysics, but metaphysics with a difference.[34] It has concealed from itself its own metaphysical truth, doubly throwing Being into oblivion. Once the world is esteemed as value, and the truth of the technical world as the will to power comes into its own, all questions concerning the worlding of value become superfluous and remain that way. If, as the Thinker argues we should, we think Nietzsche's overthrow of metaphysics as metaphysical despite itself, then it and everything it overcomes remains nihilistic. Interpreting God and all the metaphysical as the highest of all values, devaluing them, and interpreting them as the will to power, Nietzsche's metaphysic is not thought from out of Being itself. Taking the will to power as the Being of everything, it unknowingly gives God, the first of all beings, the ultimate blow by degrading God to the highest of all values. God died not because he could not be believed or because his existence could not be proved but because he becomes the highest being, the first cause, and then the highest value, a cultural and social artifact.[35]

This attack against God comes not first and foremost from Nietzsche or from secular humanists, atheists, and their ilk but from Christian theology itself.[36] Secular humanism does only what Christianity first started when it cut Man off from the holy by moving it to a metaphysical beyond. Discoursing on the being that is of all beings most in being, Christian theology never stops to think of Being itself, to stand astonished before the actual presence of the holy in the life and dwelling of humanity. Christianity cuts Man off from the earth, the mysterious abyss from which all things come, by knowing God only through scripture. Through scripture and dogma, it continually thinks God as only *a* being, the underlying metaphysical reality of the world, never as a truth made present by the world worlding. According to Nietzsche, not only is God dead, but much worse and more terrifying, he was killed by all-too-human men. God was killed; the highest of all beings was vulnerable to Man's insurrection, Man's rebellious uprising into the self-positing I-ness of the ego cogito. Through this rebellion, which so suddenly and mysteriously was thrust on us, everything is transformed into object, and the objectivity of the object is swallowed up by self-positing subjectivity. God no longer lights up the world from the patriarchal heights of the metaphysical but now is nothing more than a value posited by the will to power, a truth to be established by experimental science. And now, even our most devout fundamentalists talk of Christian values.

According to the Thinker, a survey of our history reveals that we never have thought the truth of Being itself, a presencing thought as its own truth—not even pre-Platonic thinking, which came the closest.[37] The history of Being begins, and comes to us, with the forgetting of Being, the oblivion of that which worlds the world. The oblivion of the world worlding, then, is not the unique consequence of the age, which calls the will to power to its truth, but is due to the reality of metaphysics, which is the enduring reality of our history. This uncanny oblivion of Being, which has haunted us throughout all our history, is due to metaphysics as metaphysics. Escaping our metaphysical destiny has thus far been fruitless. If metaphysics attempts to grab hold of its truth, it does so metaphysically, thus constantly falling short of its own truth and remaining incomplete. Every time metaphysics tries to climb beyond itself, to create a metaphysics of itself, it always falls back into itself, without knowing that it has done so. Such was the fate of Marx, Freud, and especially Nietzsche, who struggled the hardest against it.

The nothing that is the ultimate truth of our technology as human power is constantly lurking in the shadows of our thought, forever threatening to break in and shatter our thoughts, our lives, our work, our dreams, because our thought remains, despite itself, a happening of truth, a revelation of earth. Everywhere, from the beginning of our history until now, nothing is befalling Being and its truth, and so strangely that the truth of Being is forever gone from us.[38] But now, in our time when modern technology has pushed Being so much further into the shadow, nihilism is attaining its completion, pushing itself to its furthest limit, revealing its character most completely. As the danger grows, so grows the possibility that truth might happen, the earth might, at last, reveal itself. When it does, we will set aside the father's fear and let the world world, the earth be.

CHAPTER SIX

A Prison *of* Freedom

> The Man described for us, whom we are invited to free, is already in himself the effect of a subjugation much more profound than himself. A "soul" inhabits him and brings him into existence, which is itself a factor in the mastery that power exercises over the body. The soul is the effect and instrument of a political anatomy; the soul is the prison of the body.
>
> —Michel Foucault, *Discipline and Punish*

MERICA, our civic religion tells us, is the land of freedom, the land that taught the world about freedom of the press, freedom of assembly, freedom of religion, and freedom of speech. And it is America that built the institutions that protected them—an independent judiciary, representatives held accountable by frequent elections, countervailing powers that keep rulers within the bounds of law. In America, as nowhere else, rulers are under control, limited in the powers that they, as individuals, have at their disposal. Because of our institutions, no one can rule in America at personal whim and thereby endanger the freedom of anyone else. We Americans know that we are freer than any other people in the world. We know that we are free to think what we want, say what we will, choose what we believe.

And that is *exactly* why we are so subjugated, why truth does not happen for us. Seeking to become masters of our life, subjects whose whim is the world's command, we unthinkingly subjugate ourselves to the discipline of Man's choice. Given what the constitution of our institutions and discursive practices imply, then make necessary because of their rationality, we believe that power is merely a means, that knowledge is not a production of power, and that truth is not penetrated by time and place.[1] In the age of Technoarchy, liberalism conceals within its knowledge of what freedom is the boundless will to power underlying the objectivity, the value-neutral system of government, and the tolerance for different worldviews that make it possible. Knowing itself as the freedom of press, assembly, and religion, it conceals from itself its truth, which is freedom through control, choice through mastery, power through subjugation. It is this innocence, this ideology of openness, that makes Americans blind to the totalitarian quest for absolute dominion that abides in their willing, their consenting, and their tolerance.[2]

Liberalism is innocent of its truth partially because it cannot think of power, or its tyranny, being separate from the individual human subject.[3] For liberals, people alone possess power and will it, not institutions, not discursive practices, not the marketplace, not judicial procedures, not even laws. Liberals, both Foucault and the Thinker would say, are blind to the true nature of power.[4] Knowing freedom only as individuals willing their will and power, as the ability of individuals to impose their will on others, America is blind to the much larger reality of power, what it produces, what it conceals. That the rulers of America are under control in no way limits power or makes it less total. On the contrary, it only throws it into the shadow, concealing its operation, for the archytecture of power is in no way the property or will of individual human beings. They do not possess it or acquire it because they are not capable of mastering it, for they are always and everywhere possessed by it and its truth before they act to master it. The power that controls and imprisons us is not located in any expressly human intention or capability but in an archytecture that we cannot choose to escape, that is there before we are, governing our destiny before we live it, thinking our thoughts before we think them, and choosing our way of life before we will it.

We Americans, because we are constituted as Americans, are caught up in the truth and power relations of our time, and no matter how much we try, we cannot, as good liberals, assert control over our Constitution and become its mastering subjects, for before we even make the effort, the act of choosing has made the self that chooses into a prison, a self governed by the archytecture of subjectivity. Just as a prison does, the American Constitution achieves its effects without anyone presiding over it, intending them.

The truth of everything under the governing archytecture of Technoarchy is its possibility for control. In this, Technoarchy seeks out its shadow, everything that is not under control, that thwarts it, escapes it, or mocks it, and subjects it to its reason. Even in their differences, their irrationality, and obscure resistance, the others of the age of reason—the insane, the criminal, the sexually deviant, and those who challenge the ruling ideology—must be named "other," disciplined, and made into a strategic affirmation of reason.[5] Around the others of reason, Technoarchy builds institutions of control, exclusion, marginalization, and difference. More real than anything else, Technoarchy builds a world appropriate to its truth. Among the most important of these things that Technoarchy brings forth are institutions that produce human subjects.

The truth of freedom under Technoarchy is control, human control. But so much are the institutions and discourses of control articulated, extended, and built that the subject that controls disappears into what is controlled. Eventually, the whole extent of human being becomes the pure objectivity of power relations. Seeking control over everything, Technoarchy interprets everything as power, even, at last in Foucault's discourse, human subjectivity. The subject becomes the object of a subjugation more profound than himself.[6]

According to Michel Foucault, prisons are not merely repressive mechanisms, simple juridical matters, but complex technologies for the control, subjugation, and production of Man as subject.[7] And its technology of control is not confined to the limits of the prison walls because its truth and archytecture is not; it is spread throughout the whole of our civilization, penetrating our factories, schools, medical clinics, armed forces, and systems of government.[8] Within its walls, the prison reveals and practices the technologies of control that govern our world and build our institutions and buildings. As a specific archytecture of subjugation, it can be interpreted as a general technology of power in which Technoarchy holds Man as the Reserved and reveals him as subject.

But there is something ironic about Foucault's interpretation of the subject as the subjected, only a means for nonpersonal power strategies. It itself is but an articulation of Technoarchy's truth. Foucault himself describes his works as a toolbox that those who seek to resist specific power relations, such as the prison, can use. His works have no specific political agenda, no conception of the good, no specific justice that they seek.[9] They are but a means for something else, almost anything else. Like Nietzsche before him, who interpreted everything as power and as a means to something else, Foucault is a nihilist. As nihilists, they both are imprisoned within the truth of modern technology, however much they interpret its subject as the effect of a power that exceeds its grasp.[10]

Thinking everything as power, completely dissolving the subject into relations of power, Nietzsche and Foucault remove themselves as the speakers and writers of such a power to a placeless place,[11] a dwelling removed from the earth as metaphysics has always done. Unlike more conventional thinkers of metaphysics, though, they know that they are aliens to the earth. Dissolving the subject itself into an archytecture of power relations, they have destroyed the only dwelling place of humanity in this age of technology. Amid our institutions of technology, our discourses, and our truths—everything that is near at hand—the only way to be, to actually live, in our time is as a subject. To live and act in modern society, to use the tools that it provides for life, the means that it has for providing food, shelter, transportation, health, communication, education, political space, and entertainment, is daily, moment by moment, to invoke the archytecture that has made it possible—human subjectivity. As we saw before, a tool is for something. It has its use as a tool only in the context of other tools. Using any tool as a means invokes the entire world in which it is useful.

With the dissolution of the archytecture of torture in the Middle Ages, a radically different archytecture of truth came into being in the modern age.[12] It was supported by a whole variety of technologies and institutions. According to Lewis Mumford, the development of the technology of glass production was a crucial development for the development of the archytecture of the modern self.[13] By the end of the seventeenth century, glass became a common substitute for the wooden shutter, or for oiled paper and muslin. It furthermore had become much more clear and colorless, moving beyond its former uses for medieval church decoration and becoming a transparent medium through which the world could be represented. Glass helped put the world in a frame, transforming it into an object.[14] It made possible the eyeglass, the framed window, the telescope, and the microscope. By making it possible to control the visible presentation of things, to move them near or far away, to separate one thing from another for experimental purposes in chemistry and yet see both, it made it possible for Man to see entirely new objects—the moons around Jupiter, the microbes in a drop of water, and the written word when his eyes grew old and tired.

Since it opened up the interior of the household to a new visibility, it revealed dirt where it had never been seen before—in the corners, on the covers of things, underneath the furniture—and, pushing back the shadows, it created a new standard of cleanliness. To be used to its fullest extent, glass must be clean to be seen through. It also is easy to see the slightest trace of dirt on its hard and smooth surface. Glass reveals matter out of place and helps create the necessity of putting it where it belongs—outside the frame that reveals it.

But more important than its effect on hygiene, is the effect glass, as a presence revealing a new visibility, had on the archytecture of the self.[15] Even in the times of the wealthy Roman Empire, mirrors were uncommon and not very good. The images were distorted, and the background was dark. By the sixteenth century, however, the technology of glassmaking could make excellent large mirrors that accurately represented the world, and the hand mirror became a common possession. The mirror became a metaphoric presence that dominated epistemology and philosophy. For the first time it became possible to think about re-presenting the world. More than that, it provided the metaphor for self-consciousness, seeing oneself as one was. In the presence of the mirror, the ego could see itself and think about asserting control over its appearances—shaving, makeup, hair powdering, at a surface level—but more profoundly, preparing the way for the patriarch who would assert mastery over all the earth by representing it as his own.

Humane punishment, according to Foucault, relied on a whole technology of representations, pictures of punishment, which subjected the subject to its humanity, thereby making civil society correspond to its nature, the willing that willed its rationality. The first technique designed to do this was to make punishment as unarbitrary and uniform in application as possible. A perfect punishment would be transparent to the crime it punishes, mirroring both the nature of the crime itself and the remedy correcting it.[16] Such a picture presented to civil society would function as a deterrent, a lesson immediately intelligible to criminal and society. To do this the nature of the punishment must correspond exactly to the nature of the crime: those who have committed violent crimes must be subjected to physical pain; those who have acted despicably will be subjected to infamy. The more the punishment is transparent to the crime and exactly calibrated to its task, the more effective and efficient this archytecture of representation will be in deterring crime.[17]

Besides being constituted in such a way as to be a deterrent to all society, the archytecture of punitive representation must also operate on criminals themselves, preventing a repetition of their crimes and requalifying them as a juridical subject. The technique of representation designed to achieve this end was the adjustment of the punishment to the coherence of causes governing the crime in the criminal, the will determining their criminality. It would either be made painful enough so that in the calculus of pleasure and pain, crime would not be worth the pain it was sure to bring, or it would mechanistically oppose the force causing the crime, setting into motion a set of representations that restructured the economy of interests and passions in the criminal.

However, in order for these technologies of representation to be possible, a precise knowledge of the criminal and the crime had to be accumulated. Seeking this knowledge, the humane reformers of the eighteenth century sought to construct a comprehensive table of knowledge in which each crime and its appropriate punishment would find its exact place reflected in a code of law. Once the various species of criminals had been made into an archytecture of knowledge, classified, and their crimes categorized, it was clear the same punishment could have substantially different effects on criminals from different social groups or with different character structures. The technology of reconstituting juridical subjects demanded ever-greater individualization, objective knowledge of the criminal, and precise application of the punishment to attain the desired effect. Through the archytecture of the criminal, all of society became an object for the emerging social sciences. The subject gains her mastery over herself and attains her proper place in civil society only when the social sciences have objectified her, gaining precise knowledge of the ways in which her actions are determined, her will willed, making her available for the precise application of the technology of control. Such is the ignoble origin of the social sciences.[18]

Where archytecture of torture worked with violent excess upon the bodies of the king's rebellious subjects, the archytecture of humane reform worked with calm reason on the wills of criminals. For the reformers, the body was only an avenue to the will. The aim of the humanist was no longer to crush the body, dismember it, and destroy it, producing the most exquisite agonies while delaying a hellish death, but to operate upon the will, transform the mind, and reconstitute the reason of the subject.[19]

The ideal form of punishment was not, as it would be in the next age, incarceration, but rather public works. Strung together in chain gangs, the criminal worked on roads, canals, and public squares. Traveling throughout the land, she bore the representations of her crimes, benefiting society not only with her work, which repaid the damage her crime had caused, but also, and more important, with her lesson, which demonstrated the irrationality of crime.

The signs that the convict bore did not have the physical effect of terror, nor did they bear witness to the power of the king, but rather they were a picture to be viewed, and the convict's public presence was a theater of punishment, designed and manipulated to produce good habits in the citizenry.[20] It was not terror that kept the land's peace but the gentle and humane art of imprinting pictures upon the soft fibers of the brain.

As suddenly as the archytecture of torture disappeared in the seventeenth century and was replaced by the humane power of the prison, the archytecture of the virtues that society celebrated and the vices that it hoped to eliminate by condemnation changed. According to Albert Hirschman, all the activities that surrounded moneymaking and generally denounced as vices—avarice, greed, lust for lucre and possessions—were revaluated and became honorable activities practiced by rational men. Even more radical than that, virtue itself seemed to disappear.[21]

Today in America we are no longer innocent enough to think that politicians have any virtue. In fact, the word "politician" most often comes to us as the cynical absence of virtue. Our politicians "pull strings," "play hardball," and "project public images." We all know that, despite appearances, they are doing it for themselves or, if they are a little more forward looking than most politicians, for their place in the history books. And we are not even cynical about our businessmen as they seek their interest. They are simply obeying the laws of human nature, however ugly they may be. Any sins our businessmen commit are instantly forgiven because their industry is so useful. The reason for this change, according to Hirschman, is that the pursuit of money suddenly emerged as an instrument of control.[22] In exactly the same way as the criminal's character was operated on to mirror Man's subjectivity and reclaim it, the character of everyone else in society was operated on by the marketplace.

Suddenly, in the seventeenth century people were no longer known or understood as sinners but as passionate beings, driven by powerful and often contradictory desires. Because of the compelling reality of these passions, the life of Man was too often, in the words of Thomas Hobbes, "nasty, brutal, and short." The age that began with the collapse of the medieval world and the disappearance of virtue, prided itself on seeing Man "as he really is," accepting as the truth of Man whatever base material motives propelled him.[23] The archytecture for understanding Man as a coherence of forces, a simple material being identical with his body, was prepared when he was understood as having a mind separate from his body, a spiritual being that was above his passionate being. When the world was being represented as a machine, Man's soul had to be above it, beyond it, mastering it all.

Just as mathematics, celestial mechanics, physics, and chemistry were producing laws of nature, the appearance of human passions became the possibility of formulating laws, so-called natural laws, of human motion. This explains why people such as Hobbes, Locke, and later Rousseau were so concerned with projecting a state of nature.[24] The state of nature in modern political discourse is an artificial projection, a utopian thought experiment, designed to reveal the laws of human motion.[25] Like Galileo's hypothetical universe, the state of nature need never have actually occurred in reality in order for it to establish true laws of human motion, justice, and legitimacy. Just as Galileo's mathematical projections were not strictly bound to empirical reality for their truth (they hypothesized a circumstance that his science had no access to: an airless, frictionless environment), the state of nature need never have actually have been. A placeless place created for the purpose of revealing the true origin of things, it is sufficient that it reveals Man as the father and master of a world that he can manage. Revealing Man's patriarchy, the strategic function of the state of nature is to establish and justify a technology of human control. The nightmare of a war of all against all is a dystopian thought experiment that removes irrelevant variables (the accidents of history) by controlling for the relevant ones (the passions that govern men's actions) and establishes a counterexample—see what happens when Man is not controlled. Like many who followed him, Hobbes was a utopian thinker, and like many who found his narrative of the world compelling, he made up for his idealism and impracticality by being a *brutal* utopian. People would resist his ideal, thwart it, subvert it, so they had to be made to fear it. Death would be given to all who refused the sovereign's will. All utopians ever since have admired Hobbes for his realism and his resourcefulness.

As the first application of the new political technology, America was designed in the image of Isaac Newton's vision of the solar system. At the same time that Thomas Hobbes taught the Founding Fathers that the state was an artifice willed by Man to save himself from the irrational war of conflicting wills, Newton showed them how exactly counterbalancing forces could be structured to create a long-enduring system.[26] Knowing that the planets were maintained in their orbits about the sun by the exact balance of gravity and orbital motion, the fathers of America could conceive of a state that could be made to endure because of an exact balance of political powers and individual wills.

Once the laws of human motion were known, they could be used to balance each other and halt the decay of republics into tyranny. The continuous harmonic motion of the planets about the sun provided an image for the Founding Fathers to emulate. As it was with mathematical physics, so too it could be for political science. America was the first state to be founded by the consent of the governed, built acknowledging the causality of the passions, and structured according to the laws of reason. It was the first government built expressly as a means for the will, and for nothing else. As such, it had to rip itself free from the irrational and unwilled traditions that thwarted the will in willing only itself, freeing it from its shadow. Cutting itself off from its past, setting aside any conception of the good, any acknowledgment of virtue, any of the traditions of nobility, America engaged itself in revolution and willed itself as "We the People."[27] The will became the originary archy of the American republic, and government only the means for securing its command.

But not without considerable dissent. Various groups of people, sometimes described as Anti-Federalists, sometimes as republicans, had an alternative vision of America. John Winthrop, one of the first Puritans to land in America, describes his "city set upon a hill": "We must delight in each other, make others' conditions our own, rejoice together, mourn together, labor and suffer together, always having before our eyes our community as members of the same body."[28] The Puritans had many faults, among them the belief that material wealth was a sign of God's approval, yet they did seek an alternative notion of freedom, perhaps describable as friendship. Winthrop criticized what he called "natural liberty," the freedom to do whatever one wants, evil as well as good. In contrast, true freedom —what he called "moral" freedom—is freedom "to that only which is good, just, and honest." Justice for him, as Robert Bellah and his associates argue, was a matter not of procedure or technique but of substance and forgiving friendship.

> When it was reported to him during an especially long and hard winter that a poor man in his neighborhood was stealing from his woodpile, Winthrop called the man into his presence and told him that because of the severity of the winter and his need, he had permission to supply himself from Winthrop's woodpile for the rest of the cold season. Thus, he said to his friends, did he effectively cure the man from stealing.[29]

Thomas Jefferson later shared the same thinking. Preferring small egalitarian republics to a large commercial state, he emphasized the role of community in cultivating virtue, friendship, and character. He feared that "our rulers will become corrupt, our people careless," if the people forget themselves "in the sole faculty of making money."[30] Both Thomas Jefferson and Thomas Paine advocated direct, participatory democracy as a way of cultivating the spirit of republicanism, civic responsibility, and moral virtue. It also was a way of checking the greed and power of the wealthy.

Alexander Hamilton and James Madison, who advocated a large commercial state, saw things quite differently. They described the small republics that Jefferson and Paine advocated as "tyrannies of the majority" and saw a large state with strong central authority as the best remedy to the passions of local demagogues and unruly mobs. While even Madison agreed that in the end the welfare of the state depended on the virtue and the good of the people, he doubted that it would arise from the community. Only national institutions, checking each other, setting watch over each other, would transform people's passions, raising the public good up over private interest.

According to Bellah and his associates, election and representation, for Madison, would be a filter that would select the most virtuous, most public spirited of people. This public aristocracy of merit would then set aside their specific interests and govern in the public good.[31] Perhaps. But that is not the way it turned out, as Bellah and his associates point out. Whatever notion of national community there was in Madison's thought (and it is at best marginal), it lost out to unrestrained individualism, self-interest, and commercial expansion. Whatever they were supposed to be, Madison's professional politicians ended up accommodating private interests rather than invoking civic virtue. This is not an accident. Madison and the rest of the Founding Fathers tied legitimacy of the order and the freedom of its citizens so tightly to the individual's consent, willing, and interest that there was simply no room for the virtues or the freedom that Winthrop and Jefferson preferred.

Freedom as truth and justice as substance lost out to freedom as choice and justice as procedure.

Because the will must be left to will, the only function of government became resolving conflicts between wills—protecting property, assuring the free exercise of speech, religion, assembly, and the press. Other than that, the government merely let the individual will go its way unfettered. Government became a purely technical activity, knowing nothing of virtue or the good, caring nothing for the past. Plato's Good, Aristotle's virtues, and Augustine's sins became subjective, private, and thus disarmed. The realm of the will's greatest freedom became the marketplace. There the will could engage in commerce with other wills freely, each will freely becoming a means for others.

The original argument for capitalism, according to Albert Hirschman, was not that it, as a technique of management, could produce more and better washing machines, cars, airplanes, computers, and whatever but that it could gently assert the power of a technology of the will over Man and, through a mastery of it, make Man master of himself, willing his will in a rational way.[32] Capitalism, like the prison, is in its truth a technology of human control, an archytecture for the government of the self. The new science of economics, like the new science of politics, could emulate the discoveries of the science of physics and make possible an enduring and harmonic system. No longer was moralistic exhortation, the example of the saints, or the threat of damnation enough: the passions, as causes like the forces of nature, must be traced back to their origins in the individual and, through a science of their force, brought under control. The passions were not to be repressed or attacked as sin but to be channeled and made harmless, if not useful and valuable.

Through the science of human passion and its control, the passions that threaten the safety of society—like ferocity, avarice, and ambition—are transformed into the nation's defense, commerce, and politics. From dangerous and threatening vices emerge the nation's welfare, strength, and virtue. Properly harnessed by an archytecture of control, human vices, as if by an invisible hand, become human utility.[33] The life of Man in a properly ordered state is much like Mephistopheles in Goethe's *Faust*, who is "a portion of that force that always wills evil and always brings forth good."

The key to such a transformation is the creation of an archytecture where the passions become interests. An interest is a rational passion that countervails an irrational passion. Like a force, a passion can be controlled only by another passion that either negates it or redirects it. For instance, paraphrasing Bacon, an immodest and forward woman can be made modest and induced to restrain herself by appealing to her vanity and self-love, persuading her that modesty is an invention of love and, most certainly, the means by which it is attained.[34] Understood as a coherence of forces, the passions are available to be used to control or balance other passions. As a passion, avarice occupied the hallowed position of the deadliest of all deadly sins toward the end of the Middle Ages. But when it became apparent that it could be used to control the other passions, it suddenly became acclaimed as commerce and given the task of holding back passions that had long been thought less reprehensible, like lust for glory. With this reversal, greed became economic interest and was understood as the passion of self-love upgraded and contained by reason.

People were rational to the extent that they pursued their self-interest; furthermore, people could be governed only through the management of their self-interest. Self-interest made people constant and predictable, formed them into a tight unity that revealed their abiding character.[35] Once there was something constant and predictable, laws of human behavior could be discovered, like the laws of nature that physics had discovered. When a person becomes self-interested, their actions become as transparent as they are predictable, almost as though they were a wholly virtuous person. If the impossible happened and the laws of natural interest were repealed, if people truly became disinterested and altruistic, they would break up the universal homogeneity of human nature and make the technology of rational government impossible. Each acting on their own altruism would do whatever their particular form of altruism led them to do, and thereby they would escape the management of their government.

For the seventeenth century, self-interest, as the underlying truth of human behavior, does not lie because it will not produce any action other than those that are intelligible to reason. Though each is pursuing his self-interest, the overall effect of everyone doing this is a positive benefit to society on all levels—moral, political, and economic. Everyone mutually gains from everyone else's self-interest, if only as a secondary and unintended result, because self-interest leads people to rational actions.[36] Rational self-interest bridles boundless ambition, making government possible, and as commercial activity, it links everyone to everyone else, so tightly that the welfare of all must be satisfied as a condition for the satisfaction of any self-interested intention. As people develop economic interests, they develop a rational investment in a strong web of interdependent relationships. If their property is to be secure, they must develop an interest in a government willing to protect it, both against domestic lawbreakers and against foreign invaders, and as a rational extension of their self-interest they will be willing to pay taxes for it.

For this reason the founders of America made property a qualification for participating in government. Property ownership was not only a demonstration of a person's reason, it subjugated them to the laws that universalized its practice through their self-interest. Furthermore, commercial interdependence across national boundaries makes war, the ancient scourge of mankind, less likely because it becomes a threat to trade and consequently to self-interest.[37]

According to the reign of truth that governed early American political thought, government derived its legitimacy, rulers their authority, and laws their justice from the consent of those governed by them. Unlike the feudal monarchies of the Middle Ages and the states of any age that preceded America, the American republic had its point of origin in the human subject, its justification in the individual control freedom made possible. The rulers of the feudal realm, according to the reigning truth of the Middle Ages, derived their legitimacy from their family's descent, which was understood as a sign from God of their authority.[38] God put the sovereign over their subjects, and their justice was God's judgment.

However, when the state of nature became an experimental counterexample upon which the legitimacy of the state was derived, the human subject displaced the authority and signification of God. Existing in a state of nature before society, the human self consents, as is its natural power to do, to the terms on which it will enter society, accepting certain terms, rejecting others, according to the archytecture of human self-interest. Once the self is understood as a coherence of forces, of countervailing passions, the only legitimate government is a government that is based on the human self as its cause and underlying reality. Mastering itself as a coherence of forces through its science of passions and its technical archytecture of interests, the self-governing self becomes its own prison, the self-governing subject that consents to its government and finds its will in it.[39]

It was the Quakers of Philadelphia, according to Foucault, who brought the prison to its fulfillment in 1790 with the opening of the Walnut Street prison.[40] As with the Dutch and English prisons, the economic imperatives put convicts to work paying for their correction. Under careful supervision, the convict's labor-time was organized as efficiently as possible, their day divided into productive segments. The moral imperative also followed the English and Dutch example, with each convict receiving moral guidance and spiritual direction.

The Quakers added some dimensions of their own, however, making the prison even more a site for the technology of control. Reversing torture's archytecture of secrecy for discovering the truth of the crime and its practice of public atrocity for punishing it, the Quakers had public trials and punished in secret behind prison walls, turning responsibility for the criminal's correction over to penal technicians, who had total control over all aspects of the convict's life. In this site of total control, these technicians of reason accumulated knowledge of the criminal and the crime, making detailed observations of the prisoner, conducting extensive interrogations, completing dossiers, and scrupulously classifying all relevant facts. They kept records of the criminal's progress under detention, taking special notice of their rebellion against the system or their acceptance of it. The aim of the new archytecture was to produce a docile body that could be trained, exercised, used, transformed, improved, and subjected to supervision. Where once it was a political imperative, punishment became a technical operation, best done free of emotion, meddling influence, and juridical intervention. It was but a politically neutral and rational means of producing the subject.

According to Thomas Dumm, the penitentiary the Quakers developed in Philadelphia both was a liberal institution and was becoming a democratic institution. "It was liberal because the entire force of its operations was to reconstruct the psychology of individual persons. It was to be democratic because the same operations applied to each individual."[41] According to Dumm, Benjamin Rush, a signer of the Declaration of Independence, friend to Thomas Jefferson and one of the leading intellectuals among the Founding Fathers, was the origin in America of the idea of the penitentiary as a "republican machine." Its operation was not just marginal but essential to the well-being of the Republic, reconstituting the moral habits that endangered the Republic. When mostly private institutions—the church, the family, the school—failed, producing a criminal, it was necessary for the government to take over. For Rush, people were basically their bodies.

> All the operations in the mind are the effects of motions previously excited in the brain, and every idea and thought appears to depend upon a motion peculiar to itself. In a sound state of mind these motions are regular, and succeed impressions upon the brain with the same certainty and uniformity that perceptions succeed impressions upon the senses of the mind.[42]

Vice was not a sin but a disease to be cured. And government's responsibility was to cure and reconstitute the moral faculty—to make the criminal back into a republican machine, a citizen obedient to the laws of the republic.

Now the object of a technical discourse, Man was appropriated into a system of rational organization that established his truth as a tool of production. His body is revealed as a machine and is subjected to rigorous analysis and to a training procedure that makes him as efficient as a machine.[43] Like a machine with tightly linked and interacting parts, the body is analytically reduced to an archytecture of units—the thumbs, the fingers, the wrist, the elbow, the shoulder, and so on. Each unit's movement is taken up separately, analyzed separately, and subjected to a calculated archytecture of training that reduces each unit to its most efficient operation and then combines everything into a smoothly functioning whole. Details are crucial; complex, vast, interdependent systems are built by attention to the smallest and most obscure details. Nothing must escape specification, reduction, analysis, and disciplinary training.[44]

Seeking to keep all details under its control, disciplinary technology relentlessly expands its archytecture across both space and time, imposing its framework on both as it uses them differently. If disciplinary technology is to work efficiently and effectively, it must work continuously, organizing time so that moment by moment the body's motions are specified more precisely and made more docile. The details of the body's motions must be chronographed because control cannot be applied occasionally or at regular intervals; it must be applied continuously, letting no moment escape its purpose.[45] The efficient operation of many bodies working in interdependent systems, such as factories and military forces, requires that every motion be standardized, the time necessary to do it precisely measured and applied universally to all the bodies working within it.

The control of space is as essential to disciplinary technology as the control of time is. Disciplinary technology produces its effects by projecting individuals into a carefully organized, specified, and enclosed space, a mathematical grid of truth prepared in advance for the individual.[46] In the hospital, the factory, the school, the prison, or the military field, an orderly grid, a ruling framework, is imposed on everything. Once established, this archytecture produces a precisely organized distribution of individuals available for surveillance. This projected distribution, following the same regime of truth present in the experiment, makes possible and facilitates the accumulation of knowledge on individuals by making their dispersion and location visible. It has the strategic advantage of identifying its others—dangerous groups or wandering individuals—and reducing them to docile and fixed categories, or at least of making possible their isolation.

In disciplinary technology the archytecture of space obeys the principle of eliminating uncontrolled spaces and partitioning everything into standardized units, quite often based on a structural principle of presences and absences. In this archytecture of space each slot in the grid is assigned a value and each individual is evaluated and disciplined according to his presence or absence. Within this organized space individuals can be placed, transformed, and observed very efficiently. For the most efficient production of individuals, it is necessary to systematically define beforehand the nature of the parts to be used, stockpile the individuals or parts that fit the definition, place them in places most appropriate to them, and set them to doing their designated function.

All waste, gaps, and free margins must be eliminated and enframed by the system.[47] This elimination of all slack increases the efficiency of the system and expands its control over individual parts by making possible their interchangeability. In many ways, the penitentiary is a utopian effort—an effort to make the whole world conform to an abstract, unsituated idea, whether it can or not. It defines the world according to its logic, imposes upon it its goal, and disciplines those (the shadow to every utopia) who do not meet its standards. Its dream is the dream of purity and innocence; its reality is discipline and resistance.

By the end of the eighteenth century, a century before Friedrich Taylor, factories were already organized by the same disciplinary techniques as the prison was. Foucault gives an example of the Oberkampf manufactory at Jouy.[48] It was divided into a series of specialized workshops separated by functions such as printing, handling, coloring, engraving, and drying. In the largest building, 110 meters long and three stories high, on the ground floor 132 tables were arranged in two rows. At each table a printer worked with an assistant. Because of this archytecture of space, all the workers could easily be put under supervision by a supervisor walking up and down the aisle between the two rows of tables. From this organization of space, the specific production of each pair of workers could easily be compared with the others. As in an experiment, any variable could be defined—strength, promptness, skill, constancy—and then could be observed, measured with exactness, organized into a hierarchy of merit, and then rewarded or punished. Spread out as an archytecture of visibility, a whole multitude of workers could be understood as individuals, and as individuals subjected to precisely calibrated control by an efficient economy of centralized supervision.

From the very beginning of large-scale industry, making it possible and assuring it its efficiency and gain, there existed a division of labor, fragmented into unique individuals and distributed across a tightly organized space and precisely measured time. In such a system, the ordering of the whole multiplicity was carried out and depended on the control, surveillance, and production of the individual. The individual was the focus of control, the part of the whole, that made the dominion of disciplinary technology possible and assured its expansion.

Under the reign of disciplinary technology, individuals are revealed and produced by the examination.[49] The specific variable the examination will reveal and the individual it will produce in a heterogeneous population are determined by the specific system and its imperatives—a system of correction will measure values of correction, a system of production will measure values of production, a system of medical care will measure values of sickness, and a system of death, like the concentration camp or the Vietnam War, will measure the body count. All systems will measure docility and internalization of the system.

Arrangement of the results of examination into a hierarchy is the key to expanding control and production. The examination individualizes according to a projected grid laid out beforehand, and its hierarchical organization of the results links the individual to the system, subjecting her to its dominion through the technology of reward and punishment. The examination and the hierarchy it makes possible combine to assure control through surveillance, efficiency through organization, order through an archytecture of space and time.

The first model of these techniques was not the prison (they attained their perfection only there) but the military camp.[50] Here total organization, total control, and total observation first assumed the forms they were to take. Foucault hints that the application of these new techniques was fostered by the introduction of the musket into warfare.[51] The musket decisively changed the nature of battle and required the introduction of new techniques of control. Instead of teaching men the art of using the bow, the sword, or the lance, instead of inculcating courage in the face of the enemy, men were now trained; drill displaced art, habit displaced courage, and systems of examination and organization displaced loyalty. And war became cruel in a new and different way. It was no longer fought between men who were either angry, loyal, or sinful; now it was fought between men who had become like automatons in systems of slaughter. The continental conquest that Genghis Khan led, all the Crusades, the Spanish conquest of America —all were incredibly cruel and all killed many, many people. But they were human in ways that the World War I battles at Verdun and Somme, or the atomic bombs on Hiroshima and Nagasaki, were not. These seemed possessed by a hatred indifferently machinelike, cold, somehow passionless, yet terrifyingly unlimited.

As a discursive technology, the examination takes many forms, rendering the individual visible in all its uniqueness. The imperative of the examination increasingly became part of the architect's design. The necessity of total visibility had to be built into buildings. For instance, at the Parisian Ecole Militaire the buildings were constructed with long halls of monastic cells, each cell a sealed compartment separating the individual from his neighbors—but with a peephole so that he could be observed.[52] In the dining rooms, the tables were neatly arranged for visibility, and the inspector's table was built higher than the others to assure it. The latrines had half-doors, but full side walls. This petty, suspicious archytecture of visibility was designed to produce healthy and vigorous bodies, obedient and docile soldiers, competent and qualified officers, and to assure that debauchery and homosexuality did not occur.

When disciplinary technology was applied to productive processes, the number of variables needing to be controlled increased as it played a more important part in the economy. Laziness, fraud, bad workmanship, sabotage, and illness became more costly and important to control as the efficiency and size of the industrial apparatus increased. Disciplinary technology took on a crucial economic function when it entered the productive process, an event that assumed decisive importance for the development of industrial technology.[53] As workers were subjected to disciplinary technology, their tasks were ever more precisely specified, simplified, and integrated into an archytecture of systematic production. They became automatons, mindlessly repeating simple tasks.

Once disciplinary technology was applied to workers, it became as easy as it was obvious and necessary to replace them with machines. And in turn, the reserve army of the unemployed that the machine produced subjected workers ever more tightly to the discipline of the labor market. Disciplinary technology and machine technology are in their truth really only extensions of each other, making men and women identical with the machines they work at, nothing more than replaceable parts in an archytecture that exceeds them. People still worked, but it was the archy, automatic and anonymous, that organized and produced things.

In order for disciplinary technology to deploy people in all their specificity at the functions most appropriate to them, to claim them for its archytecture, a normalizing standard was necessary to unify operations and extend control down to the level of microdetail necessary.[54] Assuming a global importance in the economy, disciplinary technology extended its dominion over matters too trivial and isolated to have been included in the legal web. Through a normalizing judgment, it imposed a whole micropenalty of time (lateness, absence, interruption of tasks), of activity (inattention, negligence, lack of zeal), of behavior (impoliteness, disobedience), of speech (idle chatter, insolence), of the body (incorrect attitudes, lack of cleanliness), and of sexuality (impurity, indecency). Because of the imperative for controlling the most detailed aspects of everyday behavior and for making it useful, almost anything could be subjected to an archytecture of micropenalties. Like a suspicious father, jealous of his claim, insecure about what would be revealed if he did not control everything, disciplinary technology made everything its own.

Within the domain of disciplinary technology, all behavior lies between two poles of use: the functional and the dysfunctional. Between these two poles a graduated series of steps can be imposed, quantifying and ranking deviations from the norm with objectivity and precision. By assessing individuals according to these norms, the truth of the individual's deviation is revealed and the penalty it imposes of itself is legitimated. Being an objective automatic measurement, normalizing judgment produces truth and its own legitimacy, all seemingly natural and normal.

The examination as a ritual of power is a subtle but important reversal of previous archytectures of power.[55] Previous forms of power, like torture, exercised their dominion through the visibility of power, constantly bringing it out into the open, putting it on display. Common people are kept in the shadows, an undifferentiated, unknown mass. The examination, according to Foucault, reverses these relations, bringing common individuals to the fore, concealing the powerful behind a neutral, objective, gaze. In schools, hospitals, the army, and the prison, the examination is a ritual of power that brings forth individuals, making them visible to an objective gaze and producing knowledge of them. The examination, we may say, is a technique called forth by the truth of the individual held as the Reserved in Technoarchy.

Held as the Reserved, the common individual becomes the

source of dossiers that accumulate the minute details of everyday life, a biography of banalities.[56] Previously the biography of common individuals had escaped the web of the legal system and any genre of writing, but under disciplinary technology they received meticulous and endlessly suspicious attention. In the feudal regime, individuality was most apparent for the nobility, and especially for those in succession for the crown. The more power one exercised, the more one was an individual, marked with honor, prestige, and, in death, with elaborate funerals. But in Technoarchy, a kind of writing that had once lauded heroes, celebrated the nobility, and focused on the unique and special now was reversed, and the most mundane activities accumulated in file after endless file. Clerical paperwork becomes essential to the operation of disciplinary technology, enabling the authorities to fabricate a network of objective codification. In it, the child, the patient, the criminal, and the recipient of public charity are known in much greater detail than the adult, the healthy individual, the law-abiding citizen, or the employed worker. These misfits, these problematic people, are the shadows haunting the norm, the others who have failed the utopian ideal, and to maintain the purity of the ideal, they must be known and differentiated as such—marked out, made safe by being labeled different.

The more the individual is examined and documented, and the more the documentation is systematically organized, the more disciplinary technology is able to measure the gaps between them and calculate their distribution in a given population. The archytecture of documentation makes possible the measurement of general facts, the description of groups, the characterization of collective phenomena. It is no accident that the science of statistics, which made possible the experimental control of variables projected into the population, was progressing by leaps and bounds in this age, or that the American Constitution required a census every ten years. Objectified, analyzed, and fixed within a mathematically rigorous projection, the modern individual is a historical fabrication, an archytecture unique to our age. Supporting this triumph of technology and extending it, the social sciences are made possible and necessary by its imperatives. Originating within particular institutions of power (schools, prisons, hospitals, the military), the modern social sciences (psychology, demography, statistics, criminology, social hygiene, and so on) are not mere neutral means for a humane technology of humans; rather, they are the tools of human subjugation to Technoarchy. Knowable Man is the object and product of disciplinary technology, the truth around which the social sciences deploy themselves as a mere means in the service of the human subject.

Jeremy Bentham's plan for the Panopticon, according to Foucault, is the archetype of disciplinary technology.[57] In its archytecture it brings together and assures a political technology of control. Where torture had contained with heavy chains, cold iron, and thick stone walls and had controlled with sheer terror, the Panopticon contains with light and controls with visibility. As an archytecture of power, the Panopticon consists of a large courtyard with a tower in the center; circling around it, at least part way, are a series of cells, usually several levels of them.[58] In each cell, there are two windows or openings, one bringing in light from the outside, the other facing the tower, the commanding center. The focus of this optics of visibility—the tower in the center—has large observatory windows that, protected by one-way glass or shades of some sort, can be seen out of but not into. The originary center and judge must itself remain unexamined. In this archytecture of controlled visibility, the cells become small theaters in which each actor is alone, totally individualized, and constantly visible. The individual in the cell is perfectly visible to the watcher in the tower, but only to the watcher in the tower. Separated by walls, they are cut off from contact with anyone else. Isolated from any diversion, totally visible to the authorities, they are the object of knowledge, never the subject of communication.

The major benefit of the Panopticon is its economy of organization, containing and controlling the individual not with brute force but with an archytecture of visibility.[59] Inmates in the cell cannot see if the authorities are in the tower or not, so they must always behave as if surveillance is constant, unending, and total. It is not the walls or the bars that imprison in the gentle age of reason; the prisoners themselves build the architecture that imprisons them. Through the archytecture of visibility, they internalize the disciplinary power that is operating on them. The prisoners become their own guards. So perfect is the Panopticon's archytecture that even if there is no one there watching the prisoner, they could not escape it.

The political technology of the Panopticon is continuous, disciplinary, and anonymous. Anyone, and perhaps no one, could operate it as long as it was organized properly, and anyone could be subjected to its organization. The technology is multipurpose. It could operate as easily on a criminal as it could on a madman, a worker, or a schoolboy. If the archytecture of the Panopticon functioned perfectly, it would eliminate even any thought the disciplinary archy of subjectivity could not accept. No shadows in this utopia.

The efficiency of the Panopticon does not end with its control of the individual in the cell but extends itself to the control of the controllers. Those who stand in the central tower are themselves trapped by an archytecture of visibility. Any failure on their part to control and modify the behavior of the inmates, any lack of diligence in surveillance, quickly becomes visible in the inmates' behavior to the administrators of the controllers. The accumulated knowledge of inmate behavior under the reign of disciplinary technology becomes the norm by which the controllers are judged, and any deviation from it is the mark of their failure.

Making everything its own, disciplinary technology draws everything into its archytecture, controlling inmates and controllers alike, revealing everything as object to be known and subjected to control. Seeking to make Man the master of his world, disciplinary technology escapes anyone who would master it by making their position of mastery contingent on their subjugation to its reign over them. There are no masters in the age of disciplinary technology, only prisoners of Technoarchy. The subject is a trap; the patriarch who would rule, a prisoner of his own archytecture, his technology and reason.

The Panopticon is not only a technology of control for individuals but also a laboratory for their transformation, a controlled archytecture for the execution of experiments.[60] The individual cells, the optics of visibility, the possibility for controlling all variables, and the total availability of the individual for technical control combine to make the Panopticon a perfect utopia for performing experiments on human beings. The inmates are already appropriated by archytecture of disciplinary technology, and it is their destiny to be objects of knowledge in the experiment.

Bringing together knowledge, power, control of the body, and control of space into a subtly effective political technology, the Panopticon is a particular instantiation of a general archytecture for the subjugation of Man. A republican machine, as Benjamin Rush described it, it is an adaptable technology, useful whenever there is an imperative for making individuals or populations productive, visible for control, or useful for the accumulation of knowledge. Its basic technology can be generalized and used in different organizations of space. It is a technology of enframing, organizing, and defining spatial structures and can be deployed wherever a space can be enclosed, defined, articulated, and made available for control.

As it was always intended, we are told, our Constitution is a machine for producing laws, for self-government. And as machinery of control, the aim and purpose of the Panopticon and the Constitution are the same—to transform an unruly and passionate self into a self-governing self through the effects of mechanisms of control. As a result of the balance of powers, the extent of the American state, the accountability of elections, and many other mechanisms, ambition is made to check ambition, and all participants in the system, both rulers and ruled, internalize the archytecture and act in anticipation of its countervailing powers. Just as the modern Panopticon imprisons not with thick walls and iron chains but with an archytecture of visibility that makes the self govern itself, the Constitution controls the self through the effects of carefully differentiated institutions setting watch over it.[61] Because of the technology of our government, because it has made us all into self-governing selves, irrational passions, illiberal ideas, and "petty local tyrannies of the people" are excluded from the mechanism. As a result, governance in America becomes pure technique.[62] Because of its rational structure, no one is permitted to step outside the system and think it, shadow and all, as a whole world worlding.

It is thus, according to the Declaration of Independence, the inalienable right of the self, being its own master, to revolt against any state that violates the ends for which the self contracted to enter civil society—life, liberty, and the pursuit of happiness. Because government has its origin in self-interested human consent, it is the archytecture of that human consent, the conditions of its possibility and its own necessity, that will determine the form of the legitimate government. Since the inception of our age, neither force nor accident nor the ostensible will of God can father the state—only the human subject can. No doubt this is why democracy is universally acclaimed as the only legitimate form of government. Because people are neither angels nor altruists but "by nature" passionate, contentious, and egotistic beings, they must create a technology, a way of government that will govern their passions while fostering their interests. And so in order to obtain the kind of government that everyone would consent to, certain transformations had to be brought about in the self. Seeking freedom, the self becomes a trap.

Thomas Dumm likens American liberalism to a monkey trap in Polynesia

> that works through a subtle and ingenious mechanism. To catch a monkey, one makes a hole in a coconut large enough to allow its hand to slip inside, but small enough so that when it clenches the food that baited the trap it will be unable to remove its hand. All the monkey has to do to be free is unclench its hand. If it gives up the food, it escapes the trap. But the monkey seems unable to understand.[63]

The pursuit of liberal freedom is much like this. To get the goody, to get the right to control our life, to have our choices protected by rights, we subject ourselves to the disciplines, the shadows, that accompany the power to choose and control.

Once we are trapped, we commit ourselves to reform. Those marginalized or silenced by the disciplinary order start clamoring alternately for privacy acts and protection from hate speech, for an expansion of civil rights, for affirmative-action protection, and so on. Seeking freedom as control, we only trap ourselves more by getting it.

Seeking self-government, the self has to be made self-governing, to feel the necessity of restraining itself, of moderating its irrational passions for the sake of its self-interest. To become free, the liberal self must become self-conscious, rational, intentional, unified, and consistent. Otherwise the contracts it would consent to are void, and it becomes in need of "help" to reestablish its reign over itself. Its legitimacy originating in the consent of such subjects, the archytecture of the American Constitution is remarkably like that of the Panopticon's. Like the Panopticon, there are numerous points of surveillance, minutely specified and detailed, to ensure the precise application of power to the selves that are governing to make sure that they remain self-governing—that is to say, self-conscious, rational, consistent, and accountable for themselves.

The suspicious visibility of the Electoral College, for instance, holds the president accountable to the public every four years, as does the popular election of representatives and senators. And finally, to subject everything to the suspicion of judicial review, the lifetime tenure of Supreme Court justices gives them the autonomy necessary to maintain surveillance over all the other countervailing branches of government.

Unlike the prison, which has a central point of surveillance, the Constitution has many points of surveillance, or, as it is politely said, many ways of holding the government accountable. This is necessary because the Constitution is a Panopticon not for the deviant portions of society but for the rulers of the Republic as well as the ruled. In order to keep control safe, it keeps everyone under control. In other words, tracing its legitimacy back to it origins, the Constitution that the founders fathered is itself the origin of authority, not The People. To keep The People faithful to their original intentions, their fatherhood now that they are dead, they must be subjugated to it by the automatic operations of mutually suspicious institutions. That way their patrimony continues . . .

Once the technology of the Constitution is set into place and ambition sets watch over ambition, the procedures of government automatically govern all alike. Confined within the institutions of the Constitution, held accountable by one constituency or another, forced to acknowledge the interests of all others in the presence of countervailing institutions, the ambitious and self-interested self, satisfying its ambitions in the only way available to itself, will effectively control itself as it rules other selves. Outside the archytecture of the Constitution and the effects of the Founding Father's suspicions, the ambitious self is a constant danger to the common good; inside it, however strong its ambition, the self-interested self is a promoter of it, the instrument by which passionate and base energy is turned to good effect, to a continuing commitment to the Founding Father's original intentions. In this way, America is governed by self-control, a technology of government fathered by people long dead. And so it is not The People who govern but the archytecture of control that governs. Before The People will their will, they are possessed by their willing, their self-governing selves, the archytecture the founders fathered. That is the only truth the American political process can reveal. And we continually return to it, silencing any response to our current situation by invoking the original intentions of the founders. Their will must be willed.

We faithful sons and daughters thus continue today the patrimony of our founders by expanding the ways we can make the self self-governing, that is, subject to their will. The advance of electronic technology has made it possible to extend the archytecture of visibility embodied in the Panopticon without the building. For instance, it is now possible, with sensitive electronic apparatus, to monitor the brainwaves of individuals and, by use of computer surveillance and a history of norms, to determine whether they are performing their function in the workplace or not. Any deviation from a normal working pattern sets off warnings of some sort, perhaps recording them for future reference by the authorities, and draws the worker back to her work. First applications will probably be on air traffic controllers.

The National Security Agency now randomly monitors overseas telephone calls from the United States with computers programmed to record, analyze, and kick out for human inspection any conversation in which programmed phrases or words occur.[64] As computers become more powerful and faster, and as programmers become increasingly able to subject ordinary human speech to computer content analyses, this practice will no doubt become more useful and more prevalent. Perhaps it will be, and probably already is, combined with a program to measure and analyze voice stress to see if the person is lying.

Still in an experimental phase as an instrument to monitor people on parole is an electronic radio device that is attached somewhere to a person's body in a way the person cannot remove. It sends out a signal to nearby devices, perhaps in the home, possibly throughout the neighborhood, the city, or the entire country, that records the presence or absence of the person wearing it and sends the information over the telephone to the authorities. By means of this device the authorities can know the location of the individual at any time of the day or night. No doubt this monitoring device could be eventually used to transmit information on what the individual is doing, saying, thinking, and feeling for analysis on a computer.

Perhaps such monitoring would be necessary only for the most antisocial, uncooperative characters. It may well be that the authorities would know enough of people's activities by tracking their credit-card purchases, telephone calls, TV viewing habits, and travel plans through computer networks. But even with all this developing technology of control, the project of control is not going to be able to make the world into the utopia it intends. In fact, disciplinary technology dare not succeed, completely defeating its shadow, or else it would weaken its own structure, loosening its justification for normalizing control, confusing its organization of space and time, and delegitimating its totalitarian control. Even though it seeks its destruction, reason cannot destroy its shadow because it depends upon it, needs it as other to establish its identity.

As Foucault notes, although everyone knows that the prison fails to reform, fails to make people better, it is always offered as a solution to itself. "So successful has the prison been that, after a century and a half of 'failures,' the prison still exists, producing the same results, and there is the greatest reluctance to dispense with it."[65] The true purpose of the prison is not to eliminate delinquency or criminality, he concludes, but to distribute, accommodate, and supervise illegalities, to make the delinquent into an object of knowledge and control, and thereby to give itself an identity. By producing delinquents, criminals, and perverts, the prison justifies and extends strategies of domination. It tells us what we are, and it differentiates us from what we are not. It gives us the power to control ourselves, and *that* we are most reluctant to give up.

And so, the self that would free itself to become master of the earth is first produced and then mastered by a truth that requires its subjugation. Before it is ever freed to pursue its destiny, the American soul makes itself into a prison of reason.

CHAPTER SEVEN

The Collapse *of the* Household

I remember during the fifties, the outrage with which our political leaders spoke of the forced removal of the populations of entire villages in communist countries. I also remember that at the same time, in Washington, the word on farming was get big or get out—a policy which is still in effect and which has taken an enormous toll. The only difference is that of method: the force used by the communists was military; with us it has been economic—a free market in which the freest were the richest.

—Wendell Berry, *The Unsettling of America*

FROM THE VERY BEGINNING, we in America have been homeless wanderers; our ancestors left their homes in the Old World and came to live in the New World. And when they came, they still did not stop to dwell. They farmed a piece of ground for a few years, improving it for speculative sale, then moved on west. If they displaced the traditional peoples that already dwelled there, stealing their land, desecrating their sacred places, destroying the ecology that sustained them, and, on more than one occasion, slaughtering them with genocidal intent, it was only done in pursuit of their manifest destiny, the mastery of an entire continent.

Some Americans, especially those advocating a small republic, did come and stay in one place, and they did set about caring for it and building a community of care around it. But they were always a minority, and like the Indians before them, they were later colonized by the necessity of American Empire. Once the vast majority of people in America, farmers were displaced and dispossessed by America's progress. Too soon, they became a small minority, and the nurturance and the connection to place that was a possibility for them almost vanished. Now even the farm is no longer a place to live, a dwelling place to nurture, but a unit of production. Farming is now a business, bankers assure us as they foreclose, like any other business, subject to the same requirements of profits and efficiency. In America people dare not form an attachment to the place where they live, become situated in it, because inevitably they will have to move somewhere else, perhaps because they are forced by the banker, more often because there is a job somewhere else. Today in America, most families are spread over the entire continent, seeing each other only occasionally and on holidays. Without a permanent attachment to any place, Americans too seldom care for the place where they live; often they merely use it until it is used up and move on. As a result, neighborhoods deteriorate into urban blight, toxic waste is dumped in people's backyards by organized crime, and the entire country is littered with junk.

Because farming is no longer an art of nurturance but a business, at least two bushels of prime Kansas topsoil are swept away by wind and water erosion for every bushel of corn that is grown.[1] In addition, the greenhouse effect, the depletion of the ozone layer, conversion of cropland to nonfarm uses, the waterlogging and salting of irrigation systems, falling water tables, and the diversion of irrigation water to nonfarm uses are all combining to undermine food production. If present trends continue, according to a number of experts, we have only a few decades before this country, holder of some of the richest farmland in the world, will be forced to start importing large amounts of food.[2] There is no doubt, when this happens, that Americans will be able to import food from Third World countries, as we now do in large quantities, but there is also no doubt that when we do so, we will be displacing the crops grown in the Third World. Land will be taken from subsistence farmers, export crops will displace staple crops for local consumption, and malnutrition will result. All of this can happen only because we have lost our connection to place.

The most profound effect of Technoarchy has been the separation of workplaces from dwelling places, the dissolution of the household as the primary place where things are made and its replacement by the disciplinary regime of the factory. The way of the premodern household, for all its many patriarchal faults, was a way governed by handmade care, a care that was near to the dwelling place, responsive to it and the things that surrounded it. The way of the factory, in contrast, is a way governed by necessities far from its place and global in extent. Cold, vast, built as if time did not pass for it, it is a way infinitely distant from the thing and the place that would gather it into being. Its truth is a truth that is callously indifferent to the dwelling place, the community of neighbors, the earth, and the thing, leveling every-thing out and placing it within the rationality of its global order.[3] Unsituated, unconnected to earth or neighborhood, the only truth that can happen for it is universal and unlimited utility.

As the household lost its connection to place and was caught up in a global economy, a radical reorganization of the work of the sexes took place. Women remained at home, now the homemakers, and men went out into the economy, leaving the dwelling place to make money. Radically reorganized by the archytecture of the world surrounding it, the word "home" came to be associated with a single sex—the woman, who, with the home, evoked a particular emotional tone, the warmth of the hearth, the security of a shelter against the travails of harsh industrial work. With the dissolution of the handmade care of the household, the sexual division between public action and private morality changed and took on an entirely new character, transforming women into passive irrational homebodies, while work was associated with men, reason, and commerce.[4] In their newly restructured place in the household, women became the contrast to men's aggression, domination, and control, especially of and against nature, with which women were identified. More oppressed than ever as the other of the world reason built, women became subordinated doubles of men, everything they were not. And their exploitation and submission followed the exploitation and submission of the earth. With the departure of men, the household ceased to be a place of family work and became a unit of consumption managed by women.

If women have always been abused and subordinated (at least since the Bronze Age),[5] it is only within our own time, the much-promised age of human freedom, that their subordination has become a problem. At the beginning of the modern age, just when Man was taking his place and the enlightening promise of human freedom beckoned, women were excluded from it because they became the sex that supposedly could not master it, that could not control themselves. Women were imprisoned by their sex, which excluded them from the realm of freedom and reason. Men would take care of all of that.

Before this time of human freedom, men remained at home, near the dwelling place, and though there was a sexual division of labor, the aim of work was essentially the same for both sexes: to nurture the health of the household and make it self-sufficient. Food, shelter, clothing, and most tools were produced and specifically made for the occupants of the household. The household drew everything near to itself and the care of its place, revealing the necessities of nurturing care and the responsibilities of each member of the household.

The term "husband," in an antiquated, seldom-used way, still carries with it the meaning that it used to share with "huswife"—to cultivate the land, to care for the dwelling place, the trees, plants, animals, and people in and around it for the health and well being of everything. According to the *Oxford English Dictionary*, "hus" originally meant "house," and "band" is derived from *bua* and *boa*, meaning "build," "dwell," "have a household." To husband means to be governed by care for the things of the household, to be frugal, to spare things and make sure they find their place within the household's economy. To husband, then, following the way of thinking the Thinker taught us, is to set things free, to let them be the things they are as they are in the truth, the nurturing care, of the household.

Originally, freedom is sparing; it is the care the lover shows the beloved. According to the *Oxford English Dictionary*, freedom is derived from "beloved" or "friend" in a variety of ways, some of them going all the way back to Sanskrit. It is derived from the Old English *freon*, a cognate of the Sanskrit *priya*, "dear": thus the Old English *freond* and, later, *friend*, all meaning "beloved." Throughout its past, freedom has been contrasted to slavery, a way of Being outside the love and nurturing care of the family. Slaves are not free, because they are but tools of use, a means violently set upon and constrained to its utility, while the beloved members of the household—the wife, the husband, the sons, the daughters, the near friends and relatives—are spared the trials and tribulations of a mere being as utility. As a being, a presence, and a truth, the slave is reduced to a means, a tool to be used and discarded when it is useless. In this way, as the one not loved, the slave is not free.

To practice good husbandry is to respond to the health and welfare of the beloved ones, to the things that surround them and assure them, which means keeping them safe from harm, nurturing them, and letting them be as their nature calls them to be. The slave is but a tool, a mere means evoking no such emotions. To be free, then, is to be not governed by the necessities of the instrument, to be not a means beyond the care of love, and to be spared the doom of utility. Good husbandry, good housewifery, each in its own way, is a calling to care for the dwelling place and the people that dwell there. In its truth as gathering place of human concern, the household is the place where all work is governed by care, concern, and love. Things are brought forth, present themselves, and become a gathering of concern in the household as things governed by the way of love. Truth can happen amid love and friendship in a way that it simply cannot in a global economy. This means that there are techniques appropriate for the care of the household and techniques that are inappropriate, ways that spare and husband things and ways that do not, ways that set the world free and ways that do not. The art of good husbandry and good housewifery is knowing and having at hand the appropriate ways or techniques for dwelling in peace.

According to Ruth Schwartz Cowan, the development of modern industrial technology occurred in lockstep with a corresponding development of household technology and transformation of the roles of the sexes. Contrary to popular opinion in our time, which sees the household as a sanctuary from the pressures of industrial civilization, the modern household is in fact intensely involved in it, supported, limited, and transformed by the vast systems of industrial production outside of it.[6] Quite unlike the premodern household, the modern household is tightly dependent on vast energy systems extending over entire continents for any of its "labor saving" appliances to work: it is dependent on municipal water and sewage systems for waste disposal, on agribusiness for food, on the housing industry for shelter, on the clothing industry for clothing, and on an international economy to provide it with tools, income, and social services. As the consumer which must fit into the plans of all these systems, the modern household, and the woman who manages it, is the object of much external management, usually by the advertising industry, but also through choices made necessary by the archytecture of industrial technology.

For example, according to Michael Best and William Connolly, the luxuries of one era become the necessities of the next because the archytecture of consumption on which everyone is dependent requires it. In the 1920s it was a common practice to store food in an icebox because in cities the iceman regularly came around to deliver ice. Moreover, corner grocery stores were common, making it possible to buy groceries on a daily basis and bring them home on foot. But soon the middle class started buying electric refrigerators in mass and buying groceries at supermarkets accessible only with cars. As a result, it was no longer profitable to deliver ice, and corner grocery stores were forced out of business. While the middle class could afford the change, it was a harsh burden for poorer classes. Even though they could hardly afford it, they were forced to adopt middle-class practices and tools. Even though income distribution between the classes may have remained the same, the relative ability of poorer classes to participate in the goods of society drastically deteriorated because of the introduction of new household technologies.[7] For many people it was not an uncoerced choice to change their practices but a necessity imposed by the new archytecture of technology. Cars became necessities because work and the sources of most consumption goods moved out of the neighborhood. Public transportation was excluded because transportation by car was more profitable for the transportation monopolies and because the government quickly began a massive highway program for the automobile, establishing a subsidized archytecture of consumption that soon organized the layout of cities and made other modes of transportation unavailable.[8]

In order for the household to assume its destiny as consumer for industrial factory production, many "labor saving" devices had to be introduced into the household. The true function of such devices was not so much to save woman's labor as it was to provide a market for factory-produced goods and to free men of their household responsibilities so that they could take their place in the factory. For instance, according to Cowan, the introduction of the electric and gas range and the centralized heating system with automatic feed eliminated the necessity of chopping wood, hauling it in, and hauling out the ashes—all men's work.[9] The introduction of refrigeration eliminated the need for men to handle ice, just as the development of the electric vacuum eliminated the need for men to carry out the carpets so women could beat them. And most significant of all, buying food at the grocery store, which became woman's work, eliminated much of the man's responsibility for directly producing and processing food. Factories began to make boots and shoes, so that men no longer had to work leather at home.

They also began to produce pottery and tinware, so that men no longer had to whittle containers and utensils at home. Piped household water meant that children and men no longer had to carry it in. The development of the meat-packing industry, combined with the infrastructure of the refrigerator and the transport industry, meant that men no longer had to butcher or be involved with the care of livestock.

Meanwhile, according to Cowan, the introduction of factory-milled white flour into the home greatly increased the amount of work women had to do to make bread.[10] Unlike corn bread, the main staple bread in America before industrialization, which is easily mixed and baked, wheat bread (and especially white flour bread) must be mixed thoroughly, kneaded, yeasted, left to rise, and closely watched while baking. While it reduced significantly the amount of wood men had to bring in, the introduction of the stove made it possible, and then socially necessary, for women to shift from a single-course stew or soup to a multiple-course meal requiring more hours of preparation. Manufactured cloth also served to increase woman's work.

Before industrialization, most of the materials of clothing that people wore were unwashable, being made of woven woolen goods or of leather, which were simply brushed. As Cowan makes clear, it was not women's labor that was saved with all the advances in household technology but men's; furthermore, women's labor, instead of being saved, actually increased and, more important, was made useful to the distant necessities of Technoarchy's economies.

Instead of being a dwelling place where work was guided by love for its members, the household became an instrument of consumption for factory-centered production, a means of extracting profit. The advertising industry fostered obsessive standards of cleanliness in order to create a demand for its clients, the soap manufacturers. Guilt, embarrassment, and insecurity became the means the advertisers and the advocates of factory technology used to convince housewives that they had to use factory-made infant formula, that they had to reduce the spread of infection by using paper tissues instead of reusable cloth handkerchiefs, that they could improve their children's schoolwork by sending them off after a breakfast of factory-processed cereal, and so on.[11]

Like their husbands who worked in the factories, the women managing the household were increasingly held as the Reserved for a vast system of factory production, a means for extracting profit. The introduction of household labor-saving technology became a means of breaking down the tradition of the household as an autonomous producer responding to its own needs and turning it into a consumer controlled by the technical and economic imperatives of factory production.

Once the household began to be dependent on an industrial economy, and then enslaved to its logic, it was easy to suck it further in and subject it to control by means of its relations of dependency. Until the nineteenth century, according to Cowan, most of the grains available in American households were grown and usually milled by the household. But because large commercial mills could do it so much faster and finer and could generate white flour, households began to become dependent on the flour mill to process their grains and eventually on industrial bakers to bake their bread. By 1860, according to Cowan, flour milling was the leading American industry, and the value of its product was more than twice the value of the cotton industry and three times that of iron and steel.[12] The use of factory-milled flour represents one of the first stages of the industrialized household and of the collapse of its nurturing governance as handmade care.

Because milled flour was a commercial activity, involving bankers, taxes, stockholders, factory-made tools, and extensive transportation systems, the household had to have cash in order to benefit from the flour-milling system. Bartering with neighbors and friends could not grant access to the budding industrial system because it could not supply it with what it needed: a universal medium of exchange through which it could link up with its supporting systems. The only means of access to its products was by participating in its development, either as a farmer growing a cash crop, or as a wage laborer or an entrepreneur of some sort. Instead of managing their farms as the nurturing art of husbandry calls on them to, farmers had to learn to manage their farms in a way that would fit into the industrial economy. Once farmers began to participate in the market in order to have access to the products of industrial technology, their fate was sealed.

Farmers began to seek the new techniques—factory-made implements, nursery-developed seeds, and chemical fertilizers that could increase their yield and replace farm workers, increasing their access to more of the industrial economy while paying off their old debts. As farmers increased their total production, seeking more access to the industrial economy, more farmers became superfluous to it.[13] At the time of the American Revolution, thirteen farmers were needed to support every city dweller; by the middle of the nineteenth century, only half of the population of the United States were farmers; and now, as we saw in Chapter 1, only 2.5 percent of the population of the United States are farmers.[14]

The disappearance of the family farm and the self-sufficient household and its replacement by the agribusiness corporation and the dependent household means that people are separated from their responsibility to the earth that sustains them, lost in the industrialization of the entire world. Though the process is gradual, for the most part only barely perceptible to the participants, it ends up producing radical changes. According to Berlan and Lewontin, from the early 1900s to the present day, the ratio of purchased to self-generated inputs on the farm have increased more than 500 percent, even though land, typically the largest cost of farming, was calculated as a self-generated input![15] This means that the farmer has changed from being almost self-sufficient, generating most of the inputs for production on the farm, to almost total dependence on monopoly capital for inputs. Instead of using last year's crop for seed, the farmer buys it. Instead of using mules and horses for power, which are fed with things grown on the farm, the modern farmer buys tractors, along with the fuel, lubricants, and parts necessary to keep them running. Instead of fertilizing her fields with crop rotation and manure from livestock, the modern farmer buys chemical fertilizer and controls pests with chemicals.

With each introduction of modern technology, farmers became more dependent and locked into the dynamics and rationality of vast systems of power. It started innocently enough with the self-cleaning steel plow, the early reapers of the 1830s, and then the combine powered by the stationary steam engine.[16] Farmers were seduced by the power of the new technology to reduce labor time and speed the production process. According to Berlan and Lewontin, the new combine cut labor time by a factor of eight. Though this first phase radically reduced labor time, it was limited by the immobility of the stationary steam engine. Fields still had to be plowed with horses and the crop gathered up with human labor and animal transportation.

The dependence of the farmer on the monopoly economy deepened substantially with the introduction of mobile power. The internal-combustion engine, the differential, and the pneumatic tire freed the farmer of dependence on the draft animal and the natural restraints that they imposed. Horses could work only so hard and so long, needing rest or their health would be destroyed. Not so with the machine. It labored tirelessly and consumed energy only as it worked. Because the farmer no longer needed to feed draft animals, the land that was used to sustain them could be used to grow a cash crop. Up to 28 percent more land was available for cash production, and as soon as it was, surpluses became a burden on the market, forcing more and more farmers off the land.[17]

But this decline in farm ownership has not meant a loss of employment in agribusiness. On the contrary, for every farm that was lost, jobs were created for workers who manufactured, serviced, supplied, repaired, transported, and transformed farm inputs and farm outputs. According to Berlan and Lewontin, 50 percent of the average value added in agribusiness is added after the product leaves the farm. Moreover, another 40 percent of the average value added in agribusiness is explained by the cost of farm inputs. Farming itself adds, on average, only 10 percent of the total value of agribusiness production.[18] And so, though the number of farms has radically declined, the size of the food system has not.[19] Instead of working on farms, where they might be free to respond to the earth and nurture the household, people are now subjected to the infinitely exacting reign of the factory. The center of food production and the place of the household is no longer the farm but the city.

According to Cowan, as each generation of fathers became increasingly involved in the industrial economy, ceasing to cut, haul, and split wood, to butcher animals, to build houses, and to care for crops, each generation of sons knew less and less about how it should be done—and more and more about finding a job that paid wages in the industrial economy. Finding their work undiminished and, indeed, often increased, each generation of mothers continued to train their daughters in the pursuits of an industrialized domesticity.[20]

Industrialization has eliminated the traditional male responsibilities in the household, leaving the female ones intact, if restructured as the management of consumption. This is largely the reason men were much more likely to join the labor force than women. Freed of their responsibilities as husbands, they were "free" to do it; women, made useful as consumers, were not. As a result, the household has become the reserve of women who have not yet been "freed" by means of modern technology to join their fathers, brothers, and husbands in the factories. Inevitably, it has become the great necessity of our age to "free" women as well, to liberate them from the vestiges of the tradition of nurturance that now oppresses them and keeps them from participating in the world of reason that men have built. Women are subjects too, able to control, choose, reason, and maintain hierarchy as well as any man.

All too true. And no doubt this is the reason some feminists persist in seeing advancing technology as the means to women's freedom. Modern medical technology must be made to provide them with the means to escape the tyranny of their bodies with birth control (and perhaps with artificial insemination); industrial technology will provide them with factory-prepared meals, disposable diapers, clothes that need no ironing, surfaces that are easy to clean, and so on. At long last, technology will do what it promised to do: give women their means of escape from the nurturing responsibilities of the household, just as it gave men their escape earlier from time-consuming, arduous tasks.

When this happens, the triumph of industrial technology will be complete over the earth. Freed from modernity's concept of femininity, freed from the oppression of the remaining vestiges of nurturance, women will become, like men have already become, workers totally held as the Reserved by the machine, slaves to a technology that now requires their "liberation." With woman's final liberation, the household, as a dwelling place governed by love, will be completely destroyed, replaced by the totalitarian organization of the factory, the school, the government institution, and the prison. With the liberation of men from the responsibilities of the household by modern technology, the American household is increasingly becoming a single mother raising her children in poverty, supported by stingy, suspicious, and disciplinary government agencies. Perhaps this kind of household will soon become the norm, no doubt despite attempts by "conservative" politicians to save it from its fate by managing it properly and removing the welfare incentives that are "breaking poor families up."

If it does, we should not be surprised. Once men were freed of their nurturing responsibilities within the household, becoming nothing more than wage earners supporting it from the outside, they were left with only their emotional responsibilities. In the time of the factory, emotional responsibilities bring a heavy economic penalty, being an irrational investment of time and money. No longer productive members of a self-sufficient household, children were now an economic liability, needing large amounts of money for food, clothing, transportation, education, health, and recreation. And a wife who does not have a job may not be much less of a penalty, simply another consumer to support. Since men are treated like machines at their jobs, nothing more than tools of production, it is not likely that they will be able to develop a way of Being that treats anyone else any differently. The wife becomes simply a source of services—cook, housekeeper, nurse, and prostitute. And children become welfare recipients.

As Lillian Rubin writes of the modern working-class family man, "What happens during the day on the job colors—if it doesn't actually dictate—what happens during the evening in the living room, perhaps later in the bedroom."[21] Trapped in an endless series of jobs that leave little opportunity for self-expression, require unqualified obedience to authority, usually embody endless repetition of routines, and are at best insecure, work for these men, according to Rubin, is defined by bitterness, alienation, resignation, and boredom.[22] Without any power over their work life, humiliated and embittered by their helplessness, they assert their authority relentlessly over their wife and their children, controlling almost every significant aspect of their life—whether they get a job, who they see, what major investments the household makes. Rubin sadly observes, "On the surface, working-class women generally seem to accept and grant legitimacy to their husbands' authority, largely because they understand his need for it. If not at home, where is a man who works on an assembly line, in a warehouse, or a refinery to experience himself as a person whose words have weight, who is 'worth' listening to?"[23]

Giving up so much of herself to protect her husband's ego, the working-class housewife dismisses her own needs for self-expression and autonomy by asking herself, "What gives me any right to complain? He sacrifices so much for the family at work, and I am better off than my mother, aren't I?" And so, she desperately represses her own experiences of powerlessness, misery, alienation, and unfulfillment, attacking herself for her discontent. She stays at home and obeys her husband, living her life through him and her children. But she has her dreams, and often they take the form of getting a job like her husband. According to Rubin:

> There is, perhaps, no greater testimony to the deadening and deadly quality of the tasks of the housewife than the fact that so many women find pleasure in working at jobs that by almost any definition would be called alienated labor—low-status, low-paying,

dead-end work made up of dull, routine tasks; work that is often considered too menial for men who are less educated than these women.[24]

Caught up in the authoritarian discipline the industrial world requires, the household is not a sanctuary or a retreat from the exploitation of the workplace but rather a reflection of it. The humiliation, discipline, and hierarchy that characterize the workplace are duplicated at home, repeated in the relations between the sexes and between parents and children. The household is governed not by the care of the members for each other but by the discipline of the outer world. Unable to let its truth happen, divorce, child abuse, wife abuse, failure to pay child support payments—the ethical collapse of the family, in short—is the final result of the industrialized household.[25] When the household is no longer situated in itself, tied together by handmade care, love that is daily reflected in the things people do directly for each other, but is supported and sustained by systems of production and control that span continents, it is doomed and must disintegrate into abstract individuals, people whose being is chained to the factory as a system of production.

Unlike the household, the factory is governed entirely by a highly rational economy of means, a technology that knows only the truth of efficiency and rationality and none of the household's truth of care and nurturance. Inputs, and labor is just one among many, must be minimized against the maximum output. Reduced to a means, freed of any responsibility for nurturing care, the worker in the factory becomes a slave in the fullest, most profound sense possible. Although our age has abolished the legal structure supporting slavery, it has deepened, perfected, and universalized its practice. If the ideal of Technoarchy is universal mastery, the truth is universal slavery. As one plantation owner in the South said sardonically as he watched his workers—ragged, homeless, and impoverished—file off a bus, "We don't own slaves anymore, we rent them."[26]

CHAPTER EIGHT

Harnessing *the* Earth *to the* Slavery *of* Man

Capitalism did not create our world; the machine did.

—Jacques Ellul, *The Technological Society*

HE AGE OF MAN the subject, the positer of all value, meaning, use, and utility, is the age of the machine because a machine, in its widest sense, is a Man-made fabrication, a system of any kind, material or immaterial, brought forth by Man for his use as a means for something. Once the world is viewed as picture, a series of objects posited and organized by Man and available for his exclusive use, it becomes a coherence of forces represented in the mind of Man as a means for his willing. Everything becomes a machine—plants, animals, the world economy, the motion of the solar system, and even, most ominously, Man himself. God, so far as he, the first of all beings, is considered, becomes a mere clockmaker and the universe an elaborate clocklike mechanism, moving in a tight linkage of cause and effect toward a destiny determined from the first moment.

Once Man becomes subject, the underlying reality of all things, everything becomes an object for him, organized according to his will and utility. As an object present as Man's will, the thing is exclusively available as a way of increasing his power and assuring that the will finds only itself in its willing. It does not seem to matter much whether Technoarchy comes forth as Marxism or liberalism because, in one way or another, everything ends by becoming a means for Man, a machine for him to manipulate the world with. It is true that Communist China did place limitations on the machine, sacrificing efficiency for a higher metaphysic: social consciousness. But now under new leadership, pragmatism seems once again to have won out. Inequalities due to efficiency and the imperatives of management are returning and sometimes even encouraged. And in Sweden, it is true that some attempts have been made to humanize the factory. Instead of working at an assembly line performing repetitive tasks, workers are formed into teams and assemble, say, an entire car together.

But these various efforts to challenge the unrestrained dominion of the machine are exceptions that only prove the rule. Seeking human mastery, recognizing that the machine can dispossess Man of it, these efforts to restrain the machine and humanize the factory remain all too true to Technoarchy, for they insist only on making the machine once again subordinate to human purposes. If the machine escapes human control and enslaves Man to its logic, as Marxists argue very persuasively that it does under capitalism, then the answer is obvious: the creation of a society that can master the machine. It is, it seems, only the logic of individuals acting as individuals in the marketplace that makes the machine into a monster. Socialize the machine, subject it to the unmystified, democratic, and humane control of a society that understands it as a social instrument, and it will make possible a new and unprecedented civilization of human freedom. No longer will Man be subject to the cruel necessities of nature, but at long last, through the machine, he will become the master of nature. Marx himself was one of our age's most devout worshipers of the machine.[1] Capitalism was to be forgiven its sins because, simultaneously revolutionizing the relations of production and the forces of production, it was in the process of perfecting the machine and making possible the full attainment of humanity's reality, the being of its species.

Perhaps it is not even ironic that Frederick Taylor, the great American champion of the human machine, was celebrated by Lenin himself, who made great use of his principles of scientific management in the Soviet Union. If humanity is to come to its subjectivity through the machine, perhaps it is only after it has been sufficiently subjected to it.[2] Only after all of society—economy, class relations, social consciousness, everything—has been made into a machine, a means for revolution, can the revolution finally happen. People, or maybe only the vanguard of the proletariate, can understand the dynamics of the machine and make it into the means of universal freedom. Marxist discourse itself goes a long way toward making this happen, labeling and objectifying the structures of the economy like parts of a machine—how the interaction of the parts function to produce profit, commodity fetishes, class struggle, alienation, ideology, and so on. Dialectical materialism is, after all, the *engine* of history. (Marxists do use organic metaphors, but this difference, though being more sophisticated because more holistic, is not really a difference. The organism is still a means, and therefore a machine, however much it is a living one.)

Human subjectivity is accomplished by being the underlying cause of what occurs—the archy, or commanding origin of things. To be the underlying cause, the will that out of itself wills the first motion, the subject must set about discovering the forces that deny it its mastery, whether by means of the Marxist laws of history, the dynamics of psychoanalysis, the learned patterns of behaviorism, or any other archy or metaphysic of Man that is chosen. Once these forces have been anticipated, they can be shaped into whatever utopia the will wills. To build the modern utopia, all forms of human subjectivity, whether individual or collective, proceed by making humanity available for manipulation—that is, into a machine, if only for the moment that precedes its mastery.

Seeking human subjectivity through the collective being of humanity, Lenin no doubt concluded that the individual experience of the machine could, under the grim realities of production in the Soviet Union, be sacrificed for the collective good. It was not the individual's freedom that mattered, that made humanity into the subject (capitalism was proof of that); it was the collective's freedom, the whole of society that mattered. If collective freedom was possible only by means of subjecting the individual to the brutal regime of Taylor's factory, then so be it.

And so it goes. Although many socialists outside the Soviet Union think that the Soviet Union betrayed Marx, that it was a horrible abortion of the revolutionary process, it perhaps only revealed the full brutality of humanity appropriated as machine. Other systems may not be as brutal—they would call themselves more humane—but in their basic truth they are all the same: by making the world into a coherence of forces to be mastered, they have made it into a machine.

As an instrument of Man's command, a machine is a train of parts, of resistant bodies, connected or chained together in a certain way so that if one moves, all receive the motion of the original command. Viewed in terms of its function, a machine is a system of interdependent motions, setting action against action, force against force, in precise, predetermined ways, combining them together into a coherent whole that accomplishes the will of the subject that wills it into action. The human use of a machine is constituted, limited, and deployed by the resistances that the reaction of its parts against each other make possible. Resistance is a strategic deployment of the command within the machine, a hardness used to transmit motion and information to other parts, which act together to perform its function, its meaning.[3]

A machine becomes better able to fulfill its command, to realize its meaning, as a result of a tighter, more precise, and unambiguous connection between the parts or resistances. The machine archytect's aim is to constrain the resistances of the machine's parts, to define precisely their shape and nature, and to juxtapose them so that the possibility of any but the desired motion is eliminated, nothing but the desired information is transmitted. The more resistances are constrained to the planned motion, the more complete and useful the machine is as an instrument of command. A machine wears out, becoming ill suited to fulfill its function or perform its meaning, as the resistances wear against each other, producing slack and making possible ill-functioning and unwilled deviations from planned motions. The perfection of the machine, and its utility as instrument in the will's utopia, depends on the hardness and precision of the resistances constituting it, the unequivocality of the information they transmit. Parts must retain their shape, definition, and meaning, their strategic resistance, to fulfill their function as utility for Man,[4] to belong in the utopia he has planned out.

The more sophisticated and useful the machine is as an instrument of the command, the more essential it is that all the parts are constrained to their defined tolerances. If they are not, if there is the least bit of slack, some part somewhere in the mechanism could make an unplanned and unwilled motion. Because command wills a machine tightly linked together in a chain of action and reaction, this unplanned motion could have a disastrous effect—the more catastrophic the more complex, interdependent, rigorously linked, useful, and powerful the machine is. Either it would make the machine produce something that was not planned on, or more likely, it would cause it to destroy itself by putting its parts into unplanned configurations, like a worn-out car engine throwing its rods through its valve cover.

As Marx knew well, and yet not well enough, the decisive thing about the machine in our age of machines is not its internal structure but its relation to Man, its alleged master.[5] For Marx, the machine is an instrument of production, a tool that does the same thing that the craftsman did. They are interchangeable. Since humans are appropriated as instruments of production, the machine displaces them from their craft, subjecting them to the rule of the mechanism's utopia—so much so that what was once a craftsman becomes a mere instrument of the machine, dwarfed, misshapen, born to a life of craven toil, and her life measured and valued according to her measure and value to the machine. With an awesome assertion of its power, yet too near to us to be seen with the amazement it deserves, the machine breaks Man free of his past and draws him into its coldly rational mechanism, claiming him, body and soul, as part of its utopia.

The scientific management of the human machine in this utopia entails the division of labor—not only the social division of labor, where people specialize in certain crafts, but the detailed division of labor within the factory and the craft. The worker no longer practices her craft as a whole, as she did in the household economy, but is relegated to a narrow segment of a production process organized by the disciplinary regime of the factory.

Holding humanity as Reserved and machine, Technoarchy eventually destroys the worker's skill, her craft, and her openness to the earth by seeking to rationalize her work within the archytecture of production and to subject her to her role as its means. As operations are separated from each other and assigned to different workers, they can be analyzed in isolation and made more efficient, more consistent with the utopia scientific management is building, through an elimination of unnecessary motion. By breaking the productive process into separate parts and dissociating them from the worker's craft knowledge, it becomes possible to separate motions into parts, some that are simpler than others, and each simpler than the whole craft. And once this happens, it becomes increasingly possible to understand the worker's motions mechanically, set them within the archytecture of factory production, and replace them with a machine.[6] And even if that is not possible, a highly skilled worker can be replaced with a less-skilled worker, perhaps one even totally unskilled, making the more-skilled worker more replaceable. Building its utopia, Technoarchy degrades the craftsman into utility and reconstitutes the craft into a production process under the supervision and control of scientific management, destroying the craft as an art revealed in the hands of a craftsman and near to her life.

Building this utopia, of its own logic brings with it an archytecture of command, techniques for controlling the workplace.[7] The first technique in this utopia is to give to scientific management, the agents of reason, the responsibility for gathering together all the traditional knowledge that was possessed by the craftsman and make it rational—classifying, tabulating, and reducing this knowledge to rules, laws, and formulas. The aim is to displace the worker from her craft knowledge, freeing the production process from craft, tradition, and the worker's life. Replacing craft with reason, this technique enables scientific management to assume control of the production process, to identify norms of labor, to subject workers to it, and to make it proceed according to the projected plans.

The second general technique in this utopia, dependent on the first, is that all thinking, all craft knowledge, should be displaced from the actual production process itself and removed to a planning department where it can be organized rationally. Not only must the work be rationalized, but the workers must not have any need for reasoning in their work.[8] All reasoning must be scientific management's. Separating conception from execution, thinking from planning, this technique ensures that the worker becomes a pure means of command, alienated from any irrational care for the thing made, far removed from the reason governing the production process.

Needing a place to design its utopia, scientific management develops a need for space to keep books, records, and desks, as well as a hierarchy of planners to define quotas, make specifications, and coordinate the production process. Seeking to displace everything that is not its design, scientific management removes itself from the workplace, and in this displaced place, plans out how it is going to accomplish its designs and build its utopia. It is a testament to the necessity of planning for Technoarchy that the most ubiquitous characteristic of industrial production, both socialist and capitalist, is planning. But this is unavoidable because, needing the abstractions of reason to plan out the world, they both originate in a place far removed from the earth and the crafts of the people who dwell on it.

The third technique in this utopia is the one that gathers the first two together and applies them to production in the factory, to the actual movements of the worker amid the machinery of the factory.[9] If the first technique is the rationalization of all knowledge of the production process, and the second is the concentration of all power into the hands of the agents of reason, scientific management, the third is to use this knowledge and its concentration to build the world that reason requires, the machinery to control each step of the production process, integrating every motion into the process as a whole. This technique is where the utopia that scientific management designed is imposed on the world. As such, it is fraught with many dangers and much resistance, erupting contingencies because the earth is never in complete correspondence with the plan. Workers rebel, things break, weather frustrates, the plan is an inadequate map of reality. It takes a stern will to make the plan into a reality.

Frederick Taylor first popularized the time study as a technique to get control over the worker and displace her from her craft. A time study measures the elapse of time for each motion in a work process, normalizes the time it takes to do a motion, and then evaluates specific individuals against a hierarchy of possible time values, rewarding them or punishing them according to their utility. The prime instrument of the time study was the stopwatch.

But Taylor's time study was inadequate to scientific management's need for projection, calculation, and formal planning. Time studies could be done on an actual job only in an actual factory, not against universal standards of what the human machine was capable of. Situated in an actual place, they were not nearly utopian enough. As a result, they could not be used to plan, develop, and build more-sophisticated factories, but only to improve existing ones haphazardly after they were built. For the planning and efficient development of assembly lines, a more formal and mathematical theory of human motion was necessary, a general and abstract theory that could be universalized and then applied.

Frank B. Gilbreth, a follower of Taylor's, responded to scientific management's need for generality and abstraction with the time-and-motion study. To Taylor's time studies he added the concept of a motion study; that is, he investigated and classified the basic motions of the body independently of concrete work,[10] thus freeing them from place and situation. In a time-and-motion study, basic movements were defined and their duration was measured and broken down into units that were projected as the building blocks of any productive activity. These units of motions were called, in a variant of Gilbreth's name spelled backward, therbligs. To the stopwatch, as instrument of control, were added the chronocyclegraph, stroboscopic pictures (which were photographs of motion paths superimposed), and eventually the motion picture.

The result was a catalog of the amount of time it took the human body to do certain motions—for instance, the individual amounts of time that it took to select an object, grasp it, transport it loaded, transport it empty, and so on. Once these basic motions were cataloged, they could, as abstract building blocks, be assembled in any way necessary for work on a production line. Eliminating the need for repeated and expensive experiments and modifications, the catalog of human motions greatly assisted the planning and construction of assembly lines. No longer needing to observe any existing situations before making their plans, engineers could design the speed of the assembly line and the division of labor in stations along it all on paper. And then, like God fathering the world, they would make the world correspond with their paper utopia.

With Gilbreth's therbligs, the motions of the human body became as precise an object of definition and control as the other machines on the assembly line. And that was very precise indeed. Eventually, according to Braverman, the therblig was refined into units of 0.00001 hour, or 0.036 second.[11] With the therblig, scientific management gained control over every instant of time, coming close to leaving no shadow that was unproductive, irrational, or inefficient. They could build whatever utopia of production they could design.

As the machine evolves, asserting its metaphysic of power on things, Man evolves with it, becoming increasingly possessed by the archytecture that governs it and its destiny. As the machine draws near to its destiny as a means for the command, what is essential is not its evolving complexity, size, speed, or technical sophistication but the way in which its operations are controlled. Similarly, the application of power to such hand tools as drills, saws, grinders, wrenches, and so on need not change the nature of the handcraft; it may merely make it easier. As long as the guidance of the tool and the skill behind it remain entirely in the hands of the craftsman, whatever the power added to it, it can remain near to the life of the dweller. Only when the machine's tool is constrained to a fixed path by the machine's own apparatus and it thereby becomes available for control by some command other than the craftsman's truth (as it can be with drill presses, lathes, sewing machines, trip-hammers, and so on) does the machine begin to take on its modern utopian character.[12] But this is only a beginning. The person controlling the process can still, to some extent in some circumstances, be thought of as a craftsman, responding to the calling of the thing she is making.

Under the regime and truth of scientific management, however, the fixed-motion paths of these machines can reveal new ways to rationalize and constrain the motion of the tool, displacing it further from the hands of the craftsman and subjecting it to the utopia that management commands into being. According to Braverman, for example, a lathe can be easily automated so that it runs through a cycle by itself once a workman starts it on its way, directly transmitting management's command to the product. Once that improvement is made, it is another easy step to have the lathe automatically change tools after a cycle is complete and start itself on another cycle performing another function, and so on, in a specified sequence until the machined product is finished.[13] In these kinds of specialized machines the sequence of management's design is built directly into the machine and cannot be changed without changing the structure of the machine. The motions of the mechanism are not, as yet, so much automatic as predetermined, since the control of motion is fixed within the mechanism and has no links with external control or its own working results.

The next step in the evolution of the machine toward its utopian perfection is the introduction of automation, the control of the machine's motions with information coming from outside the direct working mechanism.[14] At a simple level, this may take the form of a feedback mechanism that measures the machine's output or regulates its motion, turning it on when its task is done or keeping its motions safely within the limits of its design. Examples of this kind of commanding could be a thermostat or a governor. At a more sophisticated level, the automated machine may measure the results of its work while it is in progress, compare the results with the design specifications, and make adjustments on itself as it proceeds, continually checking its production against the plan and adjusting itself to conform to it.[15]

With automation a limited reversal takes place. Before the introduction of automation, the evolution of the machine was from general purpose to special purpose—from, say, the hand-held drill to the single machine in a factory that simultaneously drills many holes in an engine block, mills its surfaces to final finish, taps threads where needed, and so on. Such large-scale machines could have no function other than the specific one they were designed for. The single machine that drills holes in a car block would be completely useless for a slightly bigger or different engine. These machines are made only when the continuous volume of a specific product can cover the cost of elaborate equipment. Under these circumstances production lines are very carefully designed and planned.

However, with the introduction of automation, some flexibility for commanding is regained, making the machine less limited by its situation, more able to respond to changes in plans. A lathe, for instance, can be made to do a variety of things and be controlled by reprogrammable magnetic tape rather than mechanical construction. This process of making the machine more flexible is accelerating with the development of computers, artificial intelligence systems, and robots. As the machine becomes more flexible, it becomes more an agent of reason, responding ever more surely to management's commanding. In modern car factories robots are able to spot-weld different makes of cars one after another, assemble a variety of components on different makes of cars, and spray-paint different colors on different cars, without even slowing the assembly line down. As the robot develops, assuming its destiny, it will become more "human," more able to adapt to different tasks, becoming more "intelligent" and thereby more responsive to commands.

As an assembly line becomes more automatic and rational, the separate machines along it become more adapted to each other, timed and regulated to complete their tasks in harmony with each other. Conveyors and chutes transport the various components of the product from one processing machine to another just as they are needed, becoming almost indistinguishable from the processing machines themselves. When automation reaches this point, the factory ceases to be a series of machines put in stations along an assembly line and becomes a single machine.[16] Instead of many machines performing many separate tasks, we have a single incredibly large and complex machine performing a single task. The design of the machine then ceases to focus on a part of the production process but encompasses the process as an integrated whole. Factories become, as Lewis Mumford puts it, megamachines, pure expressions of rational production.[17]

Perhaps the best example of this is General Motors' Saturn assembly system, where the entire production process, from design to the initial processing of raw materials, the fabrication of parts, their transportation and assembly on the factory floor, and the delivery of the finished product to the customer, is linked and managed by computers and computer-controlled machines. As Technoarchy centralizes control, making the production process increasingly subject to its command, it removes the worker further from the thing she makes, the governing care of her craft, and her place on earth. Originating from no place, no situation that lets the world world around it, the technology of production becomes ever more utopian, more removed, more distant, more displaced.

It is generally assumed that any increase in technology, in the potential for command of the earth, requires an increase in the skill (and the dignity?) of workers. Every advance in technology is an advance in Man's mastery of the earth. This assumption, almost a tautology in our age, is brought out every time an innovation threatens to displace more workers. Economists, our primary apologists for progress, assure one and all that although workers will be displaced from their jobs, the technology that displaces them will require jobs somewhere else that are more highly skilled, more highly paid. Indeed, it is actually in the interest of workers to support every technological advance, even if it does require dislocation, because it only secures Man's dominion over nature. It seems that anyone who is opposed to technological advance is crazy and irresponsible—a Luddite.

As long ago as the 1950s, James R. Bright of the Harvard Business School raised some doubts about this myth. After watching actual production in a variety of what were considered highly automated factories of the time, interviewing three to four hundred industrialists, and presenting tentative conclusions to a dozen or so industrial audiences, Bright concluded that in general, with some exceptions in plant maintenance, the need for skill actually decreased, sometimes becoming nonexistent, with advances in automation. In fact, for Bright the myth of skill advancing with technology was dangerous because it raised expectations, created disillusion and resentment when expectations were not met, and destroyed valid job standards by setting skill standards that were unnecessary for the job.[18]

Bright set up a "mechanization profile" of seventeen levels, each specifying a specific machine or hand function and its operating characteristics, from the most backward to the most advanced. On mechanization levels 1–4, where the tool remains in the worker's hands, Bright observed that skill was increasing with every advance in the tool. On levels 5–8, where control is constrained by the mechanism but still dependent on the worker, some skills are increasing, but most have turned downward. In levels 9–11, where the machine is partially brought under external control or automated, most skills turn downward. And in the higher levels, where automation increasingly takes control of the machine, every skill required by the worker plummets to nothing or almost nothing.[19]

More recently, Lillian Rubin has made the same observation, but with considerably more sympathy for the worker.

> Today more than yesterday—because technology has now caught up with work in the office as well as the factory—most work continues to be steadily and systematically standardized and routinized; the skills of the vast majority of workers have been degraded. So profound is the trend that generally we are unaware that the meaning of "skill" itself has been degraded as well. . . . Advancing technology means that there is less need than before for skill, more for reliability—that means workers who appear punctually and regularly, who work hard, who don't sabotage the line, and who see their own interests as identical with the welfare of the company. These are the "skills" such capital-intensive industries need.[20]

It is commonly asserted, in defense of modern technology, that while modern work is indeed increasingly deadening, the worker can escape it by going home at night. There, a more meaningful life can be pursued. But Rubin points to the interrelations between work and home.

> In fact, any five-year-old child knows when "daddy has had a bad day" at work. He comes home tired, grumpy, withdrawn, and uncommunicative. He wants to be left alone; wife and children in that moment are small comfort. When *every* working day is a "bad day," the family may even feel like the enemy at times. But for them, he may well think, he could leave the hated job, do something where he could feel human again instead of like a robot.[21]

The amount of skill workers express in their work matters—it matters for their personal self-esteem, it matters for their family's well-being, it matters for the state because it affects the nation's health, including spouse abuse rates, child abuse rates, crime rates, and suicide rates. The decisive point for all of this happens when the worker's work ceases to be a handcraft and becomes the controller and monitorer of machines, when Marx would say that the worker becomes an appendage to the machine. That is when the machine enables management increasingly to subject the worker to its reason, command, and utopia by rendering her skills increasingly unnecessary.

It really should come as no surprise that the worker loses her skill as technology progresses because the very archytecture of it dictates as much. The three techniques of the command hinge on the ability of scientific management to displace the worker from her work, transforming her into a means to its utopia, and subjecting her to its rationality. Since the worker is seized body and soul as a means for the command, she must submit to the imperative of efficiency as every other instrument does. Since skill is the antithesis of Technoarchy's command, an unquantifiable, unmeasurable, unspecifiable, and unmanageable variable, it must be eliminated as much as possible, and whatever remains subjected to disciplinary organization.

This transformation of the craftsman into degraded worker and object of Man's utility occurs most fully in large-scale production, where management is able to impose its utopia on labor. But in agriculture, where centralization and the full rationalization of production have not yet occurred, the farmer's skills have not suffered nearly so precipitous a drop. In contrast to the "skilled" laborer in the factory, who becomes a skilled worker in six weeks or at the most in six months, it takes many years for a person to develop the skills necessary to be a farmer—in fact, it is almost impossible unless one is born to it. A farmer must know how to care for the soil, how to fix equipment, how to care for livestock, and, unfortunately, how to deal with the banker and the government. A farmer likewise must know the habits of weeds and pests, the patterns of the weather, and the patterns of the market. Unlike many factory workers, who can get as good at their job as they are going to get in a couple of months' time because the tasks are so precisely specified, a farmer can always become a better farmer. There is no limit to her skill.

And yet, according to Wendell Berry, the skills of farmers too have been declining with the introduction of the machine.[22] Until the self-powering machine was introduced to the farm, each introduction of new technology brought with it an increase of skill needed by the farmer. The digging stick, the first tool of agriculture, made necessary a new skill. Instead of gathering plants where they grew by nature's power, the early farmer cultivated nature's ways and brought them forth from the earth. Cultivating a place brought with it an increasing responsibility to the earth, the seasons, and the goddess. Situated in nature, a community, and a culture, early farming was an act of worship, a cultivating of the goddess. It did not set upon the earth and demand that it submit to Man's mastery, nor did it know nature as pure utility; it brought forth the fruit of the goddess and revealed her truth.

So far as she knew how, the early farmer had a responsibility to renew and replenish the place where her family lived, to cultivate the mystery of fertility, which yearly brought forth its fruit. Under the governance of the goddess, each new advance in technology—first the use of stone implements, then the introduction of metal tools, then the use of animals—brought with it a greater responsibility because as more and more was disturbed, more and more had to be preserved. As the word "cultivation" implies (note the root "cult"), the ideas of tillage and worship are joined in culture. To till is to bring forth the goddess. These words come from an Indo-European root meaning both "revolve" and "dwell." To be human, as the word "humus" reminds us, is to take up one's place on the earth, care for it, and worship the goddess whose truth governs its fertility. These ideas are bound up with the idea of a cycle—the cycle of seasons and the cycle of life, death, and fertility that renew the earth. To cultivate is to dwell, to remain and respond to the cycle of fertility that renews the earth and sustains the household.[23]

With the introduction of the draft animal, the need for skill and the limits it responded to grew together. Now humanity had to respond to the animal in a new way, as Wendell Berry argues, not as food from the hunt, but as collaborator. Two different kinds of beings were involved now in the health of the farm and the nurturance of the household, animal and human, and the health and welfare of each required the development of new skills.[24]

According to Berry, it requires more skill to use a team of draft animals than a machine because the relationship between human and animal must be cultivated. The machine is simply used. Between human and animal there are limits of natural health that must be respected, or death destroys the collaboration. Between Man and machine there are no limits, no need for care, and no need of the skill that responds to limits. Within the possibilities of its mechanism, the machine is a pure expression of human will, starting, stopping, and functioning as its manager wills it. It requires no nurturance, no cultivation, only maintenance. Displacing all limits, it has no connection to the cycle of life on earth and therefore does not require ethical restraints on its use. In its being, it is purely at the disposal of Man's willing.[25]

Cut off from the fertility of the earth, indifferent to the cycle of the seasons, the machine, as an instrument of utopian reason, enables Man to displace himself from the complex web of the earth's replenishment and to assert his mastery over it, ruling it from the placeless place of his metaphysics. In agriculture, as culture, there is a responsibility and a necessity that gives skill its place by giving it its calling: a caring for the balance between fertility and use, between gain and continuity, for the home place. The fruit that is taken one year must not be allowed to decrease the fruit that is taken the next year. Caring for place, the farmer cares for the cycle that renews it, keeping it whole, holy, and complete, with no part of it broken. Maintaining the cycle requires skill—erosion must be prevented, the way of life for all kinds of plants, animals, and insects must be known and then cultivated so that each provides a balance to the rest, and wastes must be returned to the soil so that its fertility is renewed.[26]

But the machine, displacing the farmer from her place, does not require these nurturing skills. It enables, and then requires, Man to interrupt natural cycles and balances and take what he wills. Instead of a skill that knows the life cycle of pests and their predators and uses them to maintain a balance of health, the machine gives Man a poison that kills indiscriminately. Instead of the skill that knows how to keep the health of the soil with crop rotation and humus that has been made from waste, the machine gives Man the power to apply factory-made chemical fertilizer. Instead of the skill that knows how to use draft animals, the machine gives Man almost limitless power to till the soil at will. The machine does not just give this power either; it *insists* on it. It draws Man into its utopia by making him dependent on it and its infrastructure.

The development of the agribusiness machine has resulted in the deterioration of the farmer's skill in another way. When the machine was first introduced on the farm, it was relatively simple, and farmers quickly developed the new skill necessary for repairing it. As farm machines became more complex, farmers developed their skills as mechanics. But recently farm machines have become too complex for even the most skilled farm mechanics. As farm machines increasingly use advanced electronics, high-pressure hydraulics, high-powered diesel engines, complex traction systems, and exotic materials, repair and maintenance of farm machines must be turned over to specialists, who often need the equivalent of a college education to understand the new farm machine. On my family's ranch, we have a number of very old tractors, some of them dating back to the twenties and the thirties. Even though the companies, like Co-op and Farmall, that made them have long since gone out of business or have given up making parts for them, we still are able to use them. In fact, we do so every year for one thing or another. And it is really no big deal. They are quite simple, and repairs can be done by substitution, fancy welding, and creative mechanicking.

Not so with our new tractors. I cannot count the number of times harvesting or planting has ground to a halt because some part somewhere (sometimes costing only a couple of dollars) has broken, and we cannot fix it until we get *exactly* the right repair part. It may be made with high-alloy steel, integrated circuits, tight tolerances, or whatever, but only the right part will fit. Fearing that we may mess something up, make things worse with any tinkering, we wait patiently until it shows up and replace it. If John Deere ever went out of business, we would be out of business the next day—almost no exaggeration. And so, we *replace* things now, we do not fix them.

As the machine replaces the skills of farmers, it increasingly is able to replace them. One day soon, perhaps only 250,000 farmers will produce 90 percent of the food in this country. Between now and then at least 1 million farms will disappear.[27]

The mechanization of the farm was instrumental in the creation of the factory and the displaced worker. Without a surplus of farmers, management would not have had sufficient workers willing to submit to its disciplinary regime for its labor force. The triumph of machine technology would not have occurred unless it had set upon the farmer, broken her free of her traditional practices, and then of her land, transforming her into a displaced worker vulnerable to the necessity and rationality of the factory.[28] In a real sense the Industrial Revolution was built on the displacement of farmers and their way of life. They, almost exclusively, bore the initial sacrifices necessary to create the modern factory. But more than that, the skill of many farmers was lost. As farmers lost their places, the governance of the dwelling place, the responsibility for nurturing the earth, and the life governed by the near-at-hand were lost as well. The mechanization of the farm was the first step in a chain of events that drew everything together into a totalitarian system of production that was unresponsive to the earth, the cycle of life, and the place of humanity. No one or thing could escape it, because once they participated in it, they were made more dependent on it, chained to it, and then they were lost to it.

A skill is a knowledge of the right thing to bring forth. Connected to its situation, the world that worlds it, it is an ability to discriminate, to separate things out, to know their differences, and to put them in their places. As a sense, more intuitive than rational, of what is right, appropriate, or fitting for that place, a skill is the ability to hear and respond to the earth that governs the revealing of things. From the earth, the mystery that conceals, it draws things forth, letting the world world through them. Situated in a place and in a time, a skill is appropriate only to that place and time. In a different world, governed by a different technology or way of being, it may be entirely useless, as the skill of electric welding would be in any preindustrial age.

As the Thinker would say, a skill is the ability to hear the whisper of the world worlding and let it speak through the hands of the artisan, bringing the thing brought forth near to the dwelling place of the artisan. The deterioration of skill, as merely technical skill, is not the absolute measure of humanity's degradation in the face of the machine. Although reason requires as much as possible an elimination of skill, in the age of Technoarchy there remain some highly skilled workers. But because they are at the beck and call of utopian reason, these people are in truth no different from the most degraded of workers because their skills are not directed to dwelling but to control and exploitation.

Advertising, for instance, is a real growth industry, and it does require highly skilled people to bring the consumer under its control, and yet it is an entirely suspect skill because it holds the consumer as Reserved, utilizing her as an instrument of profit.[29] Many skills that this age finds most useful are skills for controlling other people and making them into a means for some sort of use. In the United States most new jobs are opening up in such occupations as labor relations, public relations, private and public security, correction, advertising, and political lobbying and campaigning.

As the machine evolves to meet its destiny, it is followed, step by step, with the enslavement of humanity. Under the reign of the machine, whether the worker has any skill reason can use or not, they have become a slave or, collectively, a rabble, a heap growing in ignorance, incapacity, irresponsibility, and impotent resentment.[30] Asked (but in reality forced) to do nothing more than tend to the machine, the workers become incapable of doing anything else—of being citizens of the republic, of attending to the whisper of the world worlding, of fulfilling their destiny as guardians of being. Their death comes with a whimper, when it comes, and their life is lived as if it never would. Death is the great terror of the modern slave because if the thought of its possibility is ever let in, even for a moment, it becomes the measure and truth of the slave's life, and since reason knows no meaning except the slave's utility, it finds too little to justify it.

As Hegel knew, slaves die a miserable death. This is because, fearing death and fleeing from it, slaves submit to being an instrument for another, to being for them, thereby living their life by despising themselves, believing themselves a failure. Quaking to the core of their being in the face of their death, they submit themselves to a life of necessity, allowing themselves to be ruled by something far removed from their calling.[31]

This is not, as the Thinker would describe it, a good death. The life that it gathers up is not a free one; no truth has happened in it, no friendship, sparing, caring, or love—only fear, resentment, humiliation, and pain.

CHAPTER NINE

The Vulnerable Machine

As soon as the generals and the politicos
can predict the motions of your mind,
lose it. Leave it as a sign
to mark the false trail,
the way you didn't go.

—Wendell Berry, *The Country of Marriage*

ACCORDING to *Time* magazine, the Pentagon believes that the command, control, and communications system necessary for fighting a nuclear war is so vulnerable that it could easily be decapitated, making it impossible to use America's nuclear weapons as instruments of coercion or destruction in a crisis.[1] As soon as top defense officials read such a report made by Bruce Blair, commissioned by Congress's Office of Technology Assessment, they immediately upgraded it to a supersecret clearance level. So high is its classification that only the president and a few top defense officials are now permitted to see the paper. Even the author of it, who has a top-level security clearance himself, is not permitted to see it. According to one senior military officer: "This is the single most dangerous document I have ever seen." The Pentagon hastily sent an official with a top security clearance to round up all the stray copies and destroy them in a high-security incinerator in the offices of the joint chiefs.

The Pentagon's panicky attempt to maintain the secret of its vulnerability does nothing to change the fact of it. Technoarchy's power systems—military, political, economic, social, and energy—are inherently insecure and vulnerable because they create vast, tightly linked, and highly organized systems of information and power and subject them to centralized command and control. As postmodernism, especially deconstructionism, has shown time and again, such systems, unable to tolerate ambiguity, slack, or disorder, are especially vulnerable to subversive strategies, to having their meanings and functions inverted, reverted, diverted, and perverted by being recontextualized.[2] No doubt that is why we spend so much money on defense and security.

All the most sophisticated weapons in America's arsenal are dependent on the integrity of an extensive military support system and even more on a functioning industrial economy. Without a continuous supply not only of exotic parts, fabricated only in a very centralized, complex, and interdependent industrial economy, but of vast amounts of exotic fuels and more mundane things like food and clothing, America's vaunted war machine would grind to a screeching halt, a giant crushed by its own strength, rationality, and technical sophistication. Some idea of the extent of the support system necessary to put our high-tech weapons into battle is contained in the fact that it takes up to fifty technicians and service personnel to keep one F-15 or F-14 fighter ready for battle.[3]

It is a strange irony of the modern age that it is the strong, the technologically powerful, that are the most vulnerable. Despite the most intensive and most accurate bombing campaigns in the history of warfare, North Vietnam was able to continue its war effort against the United States and, even more, eventually make the most powerful and technologically advanced nation in the history of the world simply give up. North Vietnam was able to survive this onslaught because its economy was not centralized, dependent on large-scale power systems or sophisticated factory production. The bombing campaign against North Vietnam has often been compared to the bombing campaigns of World War II.[4] Vietnam is supposed to have received several times more tons of bombs than were dropped by both sides in World War II. But this comparison drastically underestimates the potential effectiveness of America's bombing campaign against Vietnam, ignoring the air force's much-improved capabilities for accuracy and intelligence. A bomb is not very effective if it misses its target or if the intelligence governing target selection is faulty, and in World War II most bombs were quite ineffective. But not so in Vietnam. Because of satellite intelligence and very sophisticated targeting devices, the air force had a good chance of hitting anything that it wanted to hit the first time it tried—but it did not do any good. North Vietnam still carried on its struggle to eventual victory.

If, however, even a small portion of the bombing campaign, with all its accuracy and intelligence, that was directed against Vietnam was turned against the United States, we would be living a life much more primitive than anything that was lived in Vietnam, a life without energy, electricity, information, food, clothing, shelter, or economic exchange, because almost everything we need in our life is dependent on the functioning of vast, interdependent, and centralized systems that are easily disrupted. This country has yet to learn the most important lesson of the Vietnam War: the radical fragility of modern life and the hopeless utopianism of centralized technology.

But actually the U.S. military does understand this quite well, as it demonstrated in the war with Iraq. The choice of targeting in Iraq reveals at least as much about the United States' own vulnerabilities as it did about Iraq's. The Pentagon hit the kind of targets that it feared most for the United States itself. The first target destroyed in the war was an electrical power-generating station near an Iraqi air-defense radar site protecting Baghdad.[5] Although the first phase of the air campaign against Iraq included conventional military targets, like Iraq's air-defense systems and radar, airfields used by Iraq's eight hundred combat planes, and Iraq's thirty main SCUD missile-launching sites, the highest priority targets were Iraq's command, control, and communications systems. Without them, the Pentagon knew, Iraq's ability to retaliate and defend itself would rapidly deteriorate, especially since Saddam Hussein had centralized so much authority under himself.

Within the first few days of the war, the United States had destroyed Iraq's entire electrical power system, its twelve major petrochemical facilities, including three refineries, and its telephone system. The U.S.-led allies then quickly attacked and destroyed Iraq's nuclear reactor and an assortment of transportation hubs, roads, bridges, and railroads.[6] Eventually, they even destroyed Baghdad's water and sewage systems, making the people of the city drink river water contaminated with human feces. All of these targets had the effect of unraveling the support system sustaining Saddam Hussein's war machine. And with amazing results. One of the largest, most formidable armies in the world was brought to its knees in a matter of weeks in one of the most lopsided victories in the history of warfare. If it cannot manage dispersed economies like Vietnam's, the Pentagon clearly knows how to handle centralized systems like Iraq's. (And equally clearly, Saddam Hussein did not appreciate the vulnerabilities of his opponent, though he is probably a lot smarter now that he knows what the United States' targeting priorities are.)

There are other ways of destroying modern systems that the Pentagon knows about but did not use in Iraq (but considered very seriously). For example, the development and use of the integrated circuit for automation, information processing, and control has made the world's industrial economies radically vulnerable to the electromagnetic pulse (EMP) effect.[7] Integrated circuits are about ten million times as prone to EMP burnout as vacuum tubes, and vacuum tubes are not immune to it. Outer-space testing of nuclear explosions in the late 1950s burned out vacuum tubes throughout the Pacific. A single one-megaton nuclear warhead exploded in outer space, according to the Pentagon, can produce a pulse of electromagnetic energy that would be strong enough to damage integrated circuits over a radius possibly up to fourteen hundred miles, enough to cover most of the United States.

EMP works this way: The gamma radiation from a nuclear blast over the atmosphere interacts with the atmosphere, specifically with the electron cloud surrounding the atom's nucleus, stripping off electrons. The result is a very powerful electromagnetic field that very quickly reaches its peak intensity, about a hundred times as fast as lightning. Because it reaches peak intensity so fast, vulnerable circuits cannot be protected by, say, a lightning arrestor.

Power lines, pipelines, telephone lines, railroad tracks, instrument cabinets, and so on pick up the pulse like an antenna in proportion to their mass, focusing its energy into any unprotected circuits. An EMP blast would cause an instantaneous, simultaneous failure of all of Technoarchy's electronic systems—among them the electric grid, pipeline controls, the telephone system, the radio and TV network, the systems keeping airplanes in flight, the electronic ignition systems of most modern gasoline engines, and perhaps even nuclear command and control centers (part of the reason why the Pentagon is so worried about decapitation).

The damage might not even be limited to integrated circuits. Power lines, for instance, would collect the pulse over great distances, building up a very powerful surge that could damage insulators, transformer windings, and probably anything like a motor that was attached to the grid. In addition, a blast of EMP would very possibly cause many nuclear power plants, especially the newer ones, to melt down. They would suddenly lose the computers necessary for automated control of the system, the instruments necessary for monitoring it, and perhaps the electromechanical machines necessary to manage the system.[8] The results of a single EMP explosion over any modern economy would be an unprecedented catastrophe—a complete failure of all the systems it depends upon for energy, transportation, communication, food, and about everything else.

Possibly the most complex instruments of power ever built, nuclear power plants are the most vulnerable of all of Technoarchy's power systems, reproducing in one system all of the modern age's inherent and unreflective utopianism, making them appropriate metaphors for the entire age. They are the ultimate expression of trying to impose abstract and unsituated reason upon the earth's anarchy. They unleash incredible forces that can be controlled and kept safe only if reason can perfectly anticipate every contingency, plan for it, and contain it. Of course it cannot, which is what makes them so utopian.

Here is the situation: Even after the nuclear chain reaction has been shut down with dampening devices, the radioactive decay from the isotopes that it has created continues. The heat and radiation from this radioactive decay cannot be reduced or controlled in any way. All that can be done is to have fail-safe devices that contain the radiation and keep the decay heat from damaging the core. At shutdown, the radioactive decay heat is 6–10 percent of the heat produced at full power. The total decay heat of the fuel in the core would be enough to melt down through a solid iron pillar 10 feet in diameter and 700 feet long. And though the decay heat slackens, rapidly at first, it is enough for weeks after the shutdown to melt the hundreds of tons of nuclear fuel in the core.

Unless the decay heat is carried away by cooling devices, it builds up, generating steam, hydrogen, and carbon dioxide, and making possible a chemical reaction between parts of the core that will generate further heat. The decay heat and its effects can build up to such an extent that it ruptures containment, spewing tons of highly radioactive waste over the countryside, causing thousands, perhaps millions, of deaths, many billions of dollars of property damage, and long-term ecological damage over possibly thousands of square miles.[9] (Chernobyl was not nearly as bad as it can get because the Russians were able to contain most of the core before it escaped into the atmosphere.)

To keep meltdowns from happening, an elaborate cooling system with much redundancy is built to carry the heat away, and if that fails, a massive containment system is supposed to keep radioactive wastes from being released. Much effort is devoted to anticipating every reasonable contingency. The key is *reasonable* contingency. Only reasonable contingencies can be planned for because, in the first place, *all* possible contingencies are simply impossible to imagine; in the second, they are too expensive to prepare for. Reason cannot anticipate every contingency, map it out, prepare for it, because it is simply too far removed from it. Situated in a placeless place, a dreamworld far removed from the earth's contingency, it can impose only its technocratic utopia on the world and try to keep it secure. But reason is necessarily an inadequate map of reality. Eventually, something has to go wrong . . .

Because they are utopian, poorly prepared to meet the earth's contingency, repairing Technoarchy's power systems after they have collapsed because of war, terrorism, natural disaster, or miscalculation is made difficult, if not impossible, by their monolithic character.[10] For many of the same reasons that they are vulnerable to catastrophic failure, Technoarchy's power systems are difficult to repair, since the repair systems are not independent but often rely, if indirectly, on the same system that they are repairing. The more time that a power system is down, the greater the possibility that its failure will cascade, causing failure even in systems that have comparatively loose links.

Quick repair is essential to stop the cascading effects of system failure because sustained failure will exceed the tolerances that dependent systems have built into them, forcing them to fail as well eventually. Because Technoarchy's economies are built around the imperative for efficiency, the tolerances of most systems are designed to handle only ordinary circumstances—the normal business cycle, typical natural disasters, peacetime bottlenecks. It is an irrational expenditure of resources to prepare for anything except reasonable contingencies; it is not profitable, not appropriate. Extraordinary occurrences—war, extensive natural disaster, severe depressions, or terrorist attacks—would easily exceed planned tolerances, making quick repair as necessary as it was impossible. Only Israel has built a national power system that has any tolerance for extraordinary circumstances, such as war and terrorism. American power systems have especially low tolerances and are capable of responding only to variations internal to the system. Any disruption outside of the planned tolerances has no corresponding repair system capable of managing it.

Technoarchy's power technologies often achieve their greatest technical efficiency only by being specially designed for the specific system. Because it draws everything into it so that it can command it, linking everything to a single monolithic system, economizing by means of large scale, there is a necessity for custom making most of the components of the system. Assembly-line production is not useful because each time a power system is built, it is likely to be designed somewhat differently, a technical improvement here, a cost-cutting measure there. The result is that the components of any one power system are unique and not easily replaceable with the components of any other power system. This custom-made uniqueness would make repair much more difficult in harsh circumstances.[11]

Severe damage to large-scale power systems typically takes many months, sometimes a year or more, to repair.[12] And repair requires a complex array of supporting systems. The repairs usually require not only small tools and welders but heavy cranes, hoists, and a well-functioning parts-supply system. Transportation of things like generator rotors and larger transformers is very difficult under even the best of circumstances, requiring extremely heavy equipment and much coordination and organization.

The manufacturing capacity for replacing key components of power systems cannot be expected to deal with much more than routine demand, and certainly could not be expected to come any where near to dealing with widespread disruption and a 10-, 100-, or 1,000-fold jump in demand. Cannibalizing some systems to get other systems working is seldom possible because of severe matching problems. Widespread disruption would quickly exhaust the small pool of highly skilled technicians necessary to make such repairs. The technical complexity of modern power systems means that they require exotic materials and fabrication techniques, which will be available only if a highly interdependent industrial economy is intact, and only if exotic minerals can be imported from a number of unstable Third World countries.

Once the infrastructure supporting the power industry is disrupted, it would be very hard to reestablish it because it itself depends on the power industry for the power it needs for its exotic fabrication techniques. If a major power source fails, its interconnections with other power sources may help it to reestablish itself, assisting it with backup and restarting, but more likely it will mean that its failure is propagated throughout other systems, rippling outward until it is as total as the rationality Technoarchy imposes on all things.[13] Because the time of Technoarchy has woven so many things so tightly together, linking divergent systems technically, economically, politically, and socially, a failure in any one power system entails disruption in others, threatening everything. Radically unprepared to meet these contingencies, it is as if we live in a technological paradise where the plan is an adequate map of reality and everything always goes according to plan. What could be more utopian than our technology?

Technoarchy's technology is never near at hand, able to be governed by handmade care, but is far flung and global in extent. Its tools of power do not draw near to the dwelling place, responding directly to its needs and its contingencies, the way that solar, wind, and animal power can, but orders everything into a vast power system whose rationality is measured by its totality, not by its purpose. Mastery is the truth of Technoarchy, distant power the measure of its truth. Knowing all things as Man's utility and means to more power, Technoarchy conceals all other ways of being, shutting itself off from the mystery of the earth, imposing its rationality on everything, insisting on the appropriateness of the technocratic utopias it builds. In doing this, it produces contingency as its shadow, its other, and makes itself radically vulnerable to its eruption. Cutting itself off from the earth, repressing it, and dismissing it, it makes its utopian dreams of reason contingent on the earth's erupting only in the ways it has planned out for it.

Seeking power over the earth's contingency, Technoarchy links everything together into a tight totality, subjecting it to a center of command and control, where Man can will his will. To be adequate to the utopia it is dreaming of, the nodes and links in these systems must be precise, sharply defined, and resistant to any contingency that would make them different, because they must efficiently transmit the effects of Man's power throughout the whole system.[14] Slack, noise, or ambiguity in the system means that control will not be effectively transmitted and will not arrive at all the dispersed parts with the same command information that it had at its inception, casting a shadow over its legitimacy. Once Man makes his claim on the world, Technoarchy reveals anything that does not fit within its precisely constrained mechanism as danger, a shadow needing to be dispersed, a threat to Man's fatherhood. Because everything is so tightly linked to the origin, the slightest ambiguity or lose linkage, perhaps something like a computer virus, can cause a reversal of meanings, injecting dysfunctional commands that, transmitted throughout the tightly organized system, could put all the parts of it into fatal configurations.[15]

Because so much depends on subordination to the origin, the primary concern in the time of Technoarchy is security, banishing the shadow. Slack, ambiguity, play, friction, out-of-place matter or dirt, and nonnormalized behavior are threats—total threats because the system must be total in order for it to assume its destiny as instrument of human control. But this concern with security reveals the vulnerability of the systems that Technoarchy has brought forth. Although they are enormously powerful, god-awful powerful sometimes, they are all brittle, able to be shattered, deconstructed with only the slightest of forces that were not incorporated into their plans. When Technoarchy's systems of control fail, when the utopia it has built fails to anticipate all possible contingencies, all the interdependent relations and functions that they had sustained, maintained, and performed fall apart, cascading like dominoes falling into each other until there is nothing but ruin. Because everything has been drawn into them by the necessity of extending power and assuring the triumph of the will, the ruin will be as total as the systems of control were powerful.

In the earth's anarchy, as in Technoarchy's economies, things are interdependent, linked in complex ways that interact throughout their entire extent. For instance, the World Health Organization launched a program to control malaria-carrying mosquitoes among the inland Dayak people of Borneo with DDT.[16] The program worked, at least within its defined limits of preventing malaria. But it also caused the roofs of the Dayak people's longhouses to start falling down, made them more vulnerable to sylvatic plague, and killed off their pets. Besides killing mosquitoes, the DDT also killed a parasitic wasp that had previously controlled the caterpillars that lived by eating their thatch roofs. Furthermore, their cats started dying off because they accumulated lethal doses of DDT from eating the lizards that had eaten the poisoned mosquitoes. Without the cats, the woodland rats multiplied, and with them the fleas that carry the plague. A major outbreak was prevented only when the World Health Organization began to parachute live cats into Borneo. Manipulating one part of nature's economy quickly spread to other parts, threatening the entire economy with collapse.

Vulnerable as nature's organic economies are to outside intervention, they are not nearly as vulnerable as Technoarchy's machine economies are. This is because nature's economies are not linked together by the truth of a center, the possibility of unambiguous command, or the necessity of efficient control. Nature comes forth as anarchy. Different species in nature adapt to and coevolve with their environment, each dispersed according to their nature. They are linked together not by the necessity of transmitting precise, unambiguous commands throughout the system but by their existence as co-dwellers, participants in nature's local household. Radically unlike Technoarchy's machine-centered economies, nature's economies have room, even a necessity, for slack, ambiguity, and play, for anarchy and chaos.[17]

To a limited extent different species can occupy the same place or niche in the economy, partially duplicating the role of the previous inhabitant. Because of this anarchycal dispersion, the more complex nature's economies are, the more different participants there are in the local household to fill different roles, the more likely it is to be stable. For instance, in an economy where rabbits are the only source of food for coyotes, the populations of both species will fluctuate widely, sometimes nearing the point of extinction.

As the population of rabbits grows, the population of coyotes, finding it easy to sustain themselves, will grow also. But they will eventually grow to a point where the population of rabbits will not be able to sustain them. Competition for the declining rabbit population will grow fierce, and the coyotes will follow the rabbits to a point near extinction, declining until there are so few coyotes that rabbits can again multiply. If, however, there are alternative sources of food for the coyotes (mice or grasshoppers perhaps) and alternative predators with somewhat different tastes (bobcats maybe), each species will be able to respond to different food sources based on their availability, giving the other species the slack they need to maintain their numbers.

Similarly, if there was some way of dispersing the regions within which the coyotes and rabbits interact into subregions, separated by, say, a mountain chain, between which either animal can move with some delay and difficulty, stability also would be promoted. Slight random variations in the population dynamics between subregions enables one region to recolonize another, since the population cycle of growth and collapse would be out of step among the subregions. Even if one animal became extinct in one subregion, it could be recolonized by the other. Because nature's economies are dispersed and heterogeneous—that is, anarchycal and chaotic—they can be stable over long periods of time if left to themselves and can regenerate themselves, even when subject to considerable stress.

Technoarchy's mechanical economies, in contrast, brought

forth and organized around the archycal imperative for human command, have none of nature's anarchycal tendencies toward dispersion and heterogeneity. (Capitalism may present itself as unplanned and decentralized, as a free market—at least its advocates describe it that way—but in fact it is a highly centralized system. First of all, everything is present to it as a commodity, locked very tightly to the reign of the market. And even if there is no human being planning out the market, as there is in some versions of socialism, Man's control, read as consumer sovereignty, is supposed to reign, the center around which it spins like a natural law. Of course, it does not happen that way, but to the extent that it does not and is in fact controlled by monopoly-planning systems, it is even more centralized, more the product of Man's will.[18] Whether planned or not, capitalism does have a center that controls everything as an instrument of Man's will.)

Indeed, seeking mastery throughout its whole extent, the pure transmission of the will, dispersion and heterogeneity are obstacles for Technoarchy to overcome. Everything must be made the same, somehow drawn up into a system of control, whether it was consumer or monopoly sovereignty. To link its systems together so that it can give cause to whatever effect it seeks and accomplish the will that commands it, Technoarchy must eliminate any heterogeneity or dispersion, all ambiguity or slack, in its machine economies, drawing everything together in a tightly woven, efficiently linked network. (Under capitalism, this means variously controlling the monopolies so that they don't mess up the free market, or controlling the consumer so that they don't mess up corporate plans.) Whatever the situation or perspective, heterogeneity, dispersion, ambiguity, anarchy, and slack dampen, confuse, and disorganize the crystalline clarity of the command, threatening mastery with its nemesis—chaos. They must be eliminated and the system allowed to function.

Unlike nature's economies, which, lacking a center, are more stable the more complex they are, Technoarchy's economies, requiring submission to the center, become more unstable and more difficult to manage as they become more complex.[19] As Technoarchy's machine economies become more complex, the interactions between different parts grow many times faster, making the design for its utopia of will incredibly complex. Each new part makes possible new relations between different parts. But unlike nature's anarchycal economies, all the new relations must be subjected to the will, made rational, and organized according to plan—made controllable by being subjected to the center of command. Despite rapidly accelerating complexity, all contingencies must be known and anticipated so that the operation of the system becomes fail-safe and the security of control assured.

That which escapes its plan—the freak, the monster, insanity, irrationality, the poor-fitting part—becomes all the more dangerous to it the more extensive and rational the plan, like dirt wearing against the fine tolerances of a complex machine. Recognizing this threat to its security, seeking to maintain its control in the face of the unplanned, Technoarchy raises the complexity of its systems of control to new heights to deal with it. But because no system is complete, able to formulate all its truths within itself, this only generates more complexity needing to be managed—and more problems and more vulnerabilities.[20] Another layer of control over a complex system may perhaps solve the immediate problem, but it will add to the network, changing relations within its complex mechanism in unplanned ways, often generating more problems elsewhere, perhaps more severe.

The more Technoarchy seeks its utopia, the will willing only itself, the more it must control; the more it controls, the more that escapes its will, its utopia. Having invested so much of itself in control, Technoarchy has no choice but to attempt to gather into its will that which escaped it yet another time. The attempt to master the earth becomes a death spiral, ever tightening in around itself as more and more escapes the limits of its expanding utopia.

For instance, to use the example of one of Technoarchy's economies, before the Great Depression, the advocates of free enterprise in America knew the role of the state in the economy as largely a "night watchman," a guardian of private property, contracts, and civility. Unlike socialism, which makes everything available for self-conscious control, liberalism restrained itself, thinking that the market was a self-correcting system. Like an automatic machine, supply would closely follow demand in a competitive, privately owned market. The consumers would be sovereign, controlling supply with their demand. But then the Great Depression occurred, bankrupting many capitalists and throwing at least a fifth of the working population of the United States out of work. Obviously control had failed. Following the lead of Maynard Keynes, liberals suddenly recognized the need for "fine tuning," for management by economic technicians, because the capitalist had a contradictory relationship to the worker, messing up consumer sovereignty.[21]

On the one hand, as individuals, capitalists wanted to pay the worker as little as possible for the work they got so they could maximize their profits. On the other hand, as the sellers of what their workers made, they needed all workers, as consumers, to have a good salary so they as capitalists could sell their product to at a profitable rate. Inevitably, as individuals locked within the imperatives of a competitive marketplace, capitalists choose to keep the wages of their workers as low as possible, hoping that other capitalists would go out of business before they did. The result of all capitalists doing this was that there was not enough demand to sustain profits, the market went out of control, and a depression started. Keynesians thought that growth could be regenerated by a taxing and spending policy that redistributed, to some degree, the capitalist's surplus to workers, increasing their demand and stabilizing the capitalist's profits. As a result, it suddenly became apparent that the national economy could be controlled through the management of a national debt.

Keynesianism was an attempt, made necessary by the collapse of the household and its displacement by the market, to expand control over the economy, to master the business cycle and regularize profits and employment. As a recognition of the limits of free enterprise to correct itself, it was a recognition of the necessity of technical management on a national scale, an ideologically awkward acknowledgment that weakened liberalism's traditional separation between public and private. Throughout the 1960s Keynesianism worked well as an instrument to manage the economy, at least by the standards of mainstream economists, but it soon became evident that the economy had to be managed in another dimension. Yet more things had to be gathered up and subjected to control.

According to the arguments of traditional liberals, the marketplace is supposed to ensure that buyers, at least on average, pay for all the costs of production and usually also a small profit to the producer. The market justly and perfectly distributes all the costs and burdens of production to everyone who benefits from them. If this utopia ever was plausible, it became increasingly implausible in the late sixties and early seventies. Because of industrial development, a whole array of externalities—costs of production not controlled by the market price—became unavoidably apparent. The environment was being polluted, the worker's health was endangered in the factory, farmland was being eroded far faster than it was being rebuilt, energy was being used up at a rate below what it cost to replace it, and toxic wastes were being dumped. In all these cases, real costs of industrial development were being sloughed off onto people who either did not participate in or benefit from the market exchanges. The full costs of production were put off on future generations, communities that lived downwind or downstream from the factories, or on a population and a workforce that was increasingly vulnerable to cancer, birth defects, and ill health. The political technology of the free market was inadequate to its task.[22]

Liberals such as Charles Schultze, who recognized these externalities and failures of the market, attempted to deal with them as the Keynesian's did—with expanded technical management, more utopia. Schultze's plan for drawing the externalities that escaped the market system back into it was to have the government research the costs that escaped market control and, though they are more often qualitative in nature than quantitative, fix a price on them. The government would then tax the industry that was profiting from these externalities at a rate that would allow no profit and thus would give capitalists every incentive for ending their socially disruptive behavior. Through technical control, the public would make use of private interest, channeling and regulating it in a way that again would make market costs reflect real costs.[23] That was his utopia.

However, Schultze's plan for handling the externalities of the marketplace, if it solves anything, would more likely introduce more problems elsewhere. One of the problems, as William Connolly has argued, is that Schultze's attempt to manage externalities through technical control implicitly depends on public managers having vast reservoirs of civic virtue, and Schultze's theory of human motivation, upon which his whole theory of management depends, explicitly acknowledges that such a reservoir of civic virtue is unlikely to exist.[24] People are self-interested and are most likely to respond to their self-interest. And even if people are not primarily self-interested but in fact are capable of patriotic action, treating them as if they were exclusively self-interested and interpreting all their actions as self-interest will make it so.

We can expect, moving from Schultze's utopia to the real world, that our public managers would be prone to corruption, an externality that could poison the Republic. Perhaps they would take bribes from the industry they are supposed to be regulating, but more likely they would overlook a "little" fact here, bend an "insignificant" rule there, so that the industry they are regulating might, say, reward them with a high-paying job later on. Once the government becomes an instrument turned toward corruption, the citizenry will no longer feel bound by its rulings and will seek their interest outside it too. As a result of no one's patriotism providing a limit to self-interest, the republic will disintegrate into a corrupt and ugly ochlocracy, perhaps further into a war of all against all.

Schultze's dilemma is the dilemma of management and technical control in the age of Technoarchy—how are the controllers themselves to be controlled so that their actions keep the system they are controlling within the limits that sustain it? The answer invariably is that a higher order of technique is necessary to control what had escaped it, a procedure or apparatus that renders the actions of managers visible, normalizes them against a pattern of acceptable behavior, and inscribes them within it by means of reward and punishment.

But this raises the system of management to a higher order of complexity, creating new relationships, new systems of surveillance and control, and new feedback loops. Inevitably, these new dynamics will have results that are unplanned, not contained within the system but needing to be. The technology of management is a utopia doomed to grow, vainly attempting to bring under control everything that escapes its necessity while adding new elements that escape control, making the system more unmanageable as it becomes more in need of management.[25]

Technoarchy's doom is its necessity for continually seeking out its other, that which lies at the margins of its systems of power, outside its borders and yet limiting its control, and subjecting it to its rationality. Long ago, before Man existed, reason knew unreason as its ontological limit, accepting its mystery as God's will. Now that it is Man's instrument and not God's truth, reason knows unreason only as disease. That which once escaped reason, and thereby revealed God's judgment, is now safely contained within a classification of difference and degree, etiology and symptomology, and the mystery of unreason is only the inability, one day certain to be overcome, of extending Man's classification to all phenomena.

As Man's instrument, reason imposes crystalline projections on the visible domain of data, sharply defining boundary, category, measure, and difference according to Man's utility. Within this utopia of rigorously posited knowledge, Man the specialist takes his place, accepting his boundary and extending the power of reason within his specialty. Seeking total control within his enclosure, Man defines variables, simplifying the complexity of his domain with unambiguous categories of difference, which he then makes available for mathematical manipulation. Through the analytic procedure of the specialist, any variable can be controlled for, revealing in its differences its lawlike relation to other analytically defined variables. By means of his developing ability to manipulate variables mathematically and control their effects, the specialist is able to move from the total control of the laboratory and the experiment to build systems for Technoarchy's economies. The analytic procedure of the experiment is identical with the analytic procedure that builds Technoarchy's political, economic, industrial, and military systems. They both yield their results by simplifying the complexity of the earth to the utility of Man. As the specialist posits variables and manages their relationships, building his utopia, he closes himself off from the earth and conceals from himself the world's real complexity. Seeking total control within the simple enclosure of analytic procedure, the specialist abandons everything that escapes his system of definitions, leaving it totally out of control.[26] A pragmatist seeking escape from the tyranny of irrelevant dynamics, the specialist repudiates and ignores the complex, subtle, and mysterious patterns that cross his boundaries, recontextualizing the meaning of his system, working in subtle unknown ways against his logic. Because he cannot control everything, and yet tries to, and builds his systems as if he did, the specialist's attempt at total control in the systems he builds is doomed to end in total disorder.[27]

The anarchy of the earth will, of itself, eventually rise up and deconstruct his elaborate utopia. And this is the more certain the more rigid and exclusive his boundary, the more precise and exact his definitions, and the more total his control. Because precision, rigid exclusion, and unambiguous logic, as artificial Man-made instruments, are impositions on the world's complexity and the earth's power to come forth in unknowable ways, the abstract forms of truth that Technoarchy projects are never adequate to the earth, and yet Technoarchy builds as if they were.

CHAPTER TEN

The Monster

> I may die, but first you, my tyrant and tormentor, shall curse the sun that gazes on your misery. Beware, for I am fearless and therefore powerful. I will watch with the wiliness of a snake, that I may sting with its venom. Man, you shall repent of the injuries you inflict.
>
> —the monster to Frankenstein, in Mary Shelley's *Frankenstein*

T ABOUT the same time that Mary Shelley was writing her dystopian tale of a Man-made monster, another monster, a nonfictional one even more terrible than Mary Shelley's, was haunting the mind of Europe—the Marquis de Sade. It seems strange that one man could provoke such an extreme reaction, unless he represented something that was all too near, too true to the reality of the time. Why has Sade been so thoroughly censored? Why was he imprisoned, locked up, and isolated from the rest of society? Was it just for what he wrote? Or was it not because his moral nihilism, much like the monster in Mary Shelley's famous novel, was the ultimate threat to the thoughts, the personal identities, and the morality of his time? Was he not locked up and isolated not so much for the crimes he might have committed but for the contagion he represented, for the scandal he made possible, and the threat to civility he posed?[1]

Could it be that Sade was, and no doubt still is, the unwelcome guest of our time, a monster representing a shadow in our thought that must be repressed, denied, isolated, and silenced—but nevertheless carefully examined (most preferably by medical doctors), studied, and carefully answered?[2] A good scientist of morals cannot just ignore Sade and his dystopianism, just jail him and censor him; he must have reasons; he must examine him (if in secret) and not fail to have the most complete answer for the isolation and denial of Sade's sort of contagion. To fail to get this monster under control is the most serious failure a science of morals is capable of.[3] Sade must be made safely other, into the not-self.

The reasons for Sade's thought, the causes of his immorality, the constitution of his mind, his body, and his soul—these all must be made to speak, to render up their truth of themselves as irrationality or immorality, to confess their secret knowledge, so that Sade and the elements, whether social or organic, which made his obscene dystopianism possible can be manipulated, controlled, subjected, and then, most important, made rational.[4] Our utopia must be made safe again. Nevertheless, a terror haunts us: as we think about the ways in which our technical mastery of morality might circle back on itself, defeating its own ends and purposes, on the ways in which it itself helps create the dystopia it denies, we begin to wonder if it is not the fearsome Sade who has trapped us, if it is not possible that the more we try to deny him, the more levers and fulcrums we reach for to pry him out of our lives, the more he reveals our utopia as a dystopia, making his dream into our nightmare and revealing the secret truth of what we are.[5]

If Technoarchy, which carries with it the necessity for mastering all the earth, cannot help but father a double, a fearsome shadow of itself that it must exclude and repress, if, in building its utopia, it must create wills, modes of behavior, or groupings of people—terrorists, delinquents, a rabble of welfare dependents, sexual perverts—that must be identified, controlled, changed, and made useful as a demonstration of reason's truth, then the legitimacy, the originary archytecture, of modern technology is called into question.[6] Its patrimony becomes illegitimate. Claiming the purest of origins for itself, needing only to be itself, it is haunted by its own shadow, what it must not be but nevertheless is. If we are in fact not other than Sade, if his dystopianism and depravity are secretly our own, the more so when we deny it, repress it, and exclude it, we can hardly make any claims to the legitimacy of our thought. It has an other in it that, by our own standards, makes it illegitimate by demonstrating the fear underlying our reason, the cruelty underlying our morality, the will to power underlying our quest for freedom, the lies underlying our will to truth. The fact that Sade and others like him are, have lived, and have thought raises doubts that we are what we would like to be.

When monsters like Sade erupt out of nowhere, as tradition has always known, they appear as a warning, a divine omen of an evil yet to come. The Old French word *monstre*, according to the *Oxford English Dictionary*, is closely related to *monere*, "to warn." The appearance of the monster, then, is something marvelous, a divine portent or warning, a fearsome event. Moreover, as its root in the French word *montre* shows, the monster appears as the effect of monitoring, of watching, displaying, showing, of revealing. As a feared abomination, a divine portent of evil, the monster is put on display (the freak show is the modern variation of an ancient practice) for all to fear, take heed of, and respond to as a warning.

And yet, as we moderns are apt to forget as we attend freak shows, as if by some monstrous compulsion within, and pity the objects put on display there, the monster is more than an abortion of nature, an accident that reason makes harmless with an etiology of its being as a coherence of forces,[7] something that is simply other, bearing no relation to ourselves than its difference. It is a summons, an irruption of Being, that reveals our own monstrous truth. Driven by some hideous compulsion, there we stand before it, watching the monster, more monstrous than it by our gawking, by our act of seeing it as other than ourselves. It reveals that which called it forth, set it on stage. As a divine warning, calling on us to reflect on our own monstrousness, the monster reveals, as it conceals. The monster reveals and conceals alternately the depravity of watching the monster, of making it other than ourselves, and the fear that makes it different from ourselves, that makes us repress our identity with it.

Kept secure in our identities by our reason, our science, and our technology, we moderns, however, know no monsters because we know no evil. Fervently committed to our dreams, repressing the shadows that make them possible, we believe there can be no nightmares in our utopias. Concealed from the gaze of science by the calm assurance of reason, the monster as divine portent of monstrosity is unknown in our age—and yet this age, of all ages, is the age of the monster.[8] Because of its truth as object of utility, the world is watched, put on display, viewed as a picture, made into an other to be dominated and used, just as the monster is put on display in a freak show. But it is not the monster that is monstrous in this display of the world as picture; it is the watching, the modern way of looking upon the world as something to be dominated and controlled. The watching that reason insists upon effects a subtle transformation on the watcher. Seeing the world as an object of its will, the watcher assumes a way of Being in the world that excludes its nurturance, disallows the possibility of love of others or compassion for the unfortunate—any relation to the world that makes it into anything except a means for the will to will itself.[9] Nothing is left to break forth from the earth of itself and left to be itself; rather, it is gathered up in the tempest of Man's willing it as a means to his willing. This way of Being conceals the monster while bringing forth the monstrous.

In Frankenstein, a dystopian tale of a Man-made monster,

Mary Shelley was issuing us a warning that goes right to the heart of this time of Man-made people, a warning about utopian fathering that has been ignored as much as it has been transformed into myth.[10] *Frankenstein* is the story of a brilliant scientist who overcomes many technical obstacles, discovers the secret of life, and uses it to create a human being in his laboratory. He becomes a father, so to speak, without recourse to intercourse, motherhood, or mothering. He is patriarchy's ideal father. The being that he fathers is brought forth only as the object of his will, and it has no other being, no other origin or connection. His claim on his progeny is uncontestable. But as soon as he succeeds and his creation awakes, Frankenstein is filled with horror, and he abandons his creation to its own devices. He cannot stand even the thought of what he has brought into the world, what it reveals about the world.

Living in a hovel near a family that he spies on, the monster quickly develops his human skills for reading, speaking, and listening to a language. He discovers his unique situation as the first Man-made man from Frankenstein's notebook, which he took with him when he left the laboratory and which contains Frankenstein's final reaction of horror to his creation. Wherever he goes, whenever he encounters another human being, Frankenstein's creation is quickly made the object of violent assault, a fearsome other. Taking this otherness within himself, accepting society's judgment of him, he flees from all humanity, becoming an outcast from the human race.

Profoundly alone, he desires companionship desperately, a friend. He seeks out his creator and prevails upon him to father him a woman too. Fearing the consequences, however, Frankenstein breaks his promise. The monster then turns against his creator in revenge, killing everyone that Frankenstein loves. Once he has lost everyone he loves, Frankenstein follows the monster to the Arctic, seeking to kill him. But he dies before he can accomplish his goal. His revenge satisfied, overwhelmed by a most terrible self-disgust, the monster destroys himself in a fiery inferno at the North Pole, far removed from any human habitation.

Contrary to the myth, but very true to it, "Frankenstein" is not the name of the monster, but of the scientist who made him.[11] Cut off from society by his monstrousness, the monster has no relation to anyone who would name him. In this curious reversal and forgotten fact lies its own truth—it is not the monster that is monstrous but the making that made him, brought him forth from unconcealment. But we see in the novel that Frankenstein himself, except perhaps for some minor flaws of character, is no monster either but, as the monster himself concludes, "is the select specimen of all that is worthy of love and admiration among men." The monster comes into the world not, it would seem, as the issue of some cruel and sadistic hand which despises the world and seeks its torment but as the exact realization of the dream of its creator, the highest attainment of the most excellent of scientists. The monster was to be the scientific demonstration of the mastery of the secrets of life, of the power for bestowing animation upon lifeless matter, renewing life where death had devoted the body to corruption. It was to be the proof of a self-conscious knowledge of humanity that would be of inestimable benefit to humankind, a total mastery of nature and man's beginnings.[12] It would make it possible for the human race to become its own father—and build its own utopia.

But the moment that the experiment was complete and the creation opened its eye, demonstrating the full accomplishment of human self-consciousness, the creator, Dr. Frankenstein, was suddenly repelled by his creation.[13] It ceased to be a technical triumph and became an abomination, ugly and hideous to look at. Frankenstein fled from it in horror, hoping that death would take it back into its grave. The dream of complete self-consciousness became a nightmare, the reality of the will being able to father itself a terror. However much he now despised it, the monster that Frankenstein created was not an accident, a miscalculation, or a failure but the accomplished reality of being able to give a total representation of Man, of being able to master the coherence of forces governing the origins of Man to the point of being able to create one. It is that technical capability for self-consciousness that is so monstrous and makes the monster so horrible. If only an abortion of nature, the monster would merely have been ugly, but as a creation of Man, the full expression of his will being able to will himself, it was an abomination, the truth of humanity as an object of science and technology. The dream of the perfect utopia, once realized, is revealed as a dystopia.

Mimicking God's creation of humanity, Man's creation of Man became the effective death of God as creator. All that was holy in Man and nature was vanquished by the new Adam, becoming nothing more than the re-presentation of a human will. With the new Adam, things were as Man willed them to be and nothing more. Everything became an object for human subjectivity, even Man himself. Technology had overcome the last obstacle, the mastery of Man himself. With its triumph, Man became a slave to himself, an object of his own utility. Frankenstein was repelled by his creation because it brought with it a radical diminution of his own status as a subject and of the subjectivity of everyone he loved. People became what they controlled, something to be controlled. If, by means of his chemical manipulations, he could create human life out of inanimate matter, asserting his will over it, he had also reduced human life to inanimate material, a coherence of forces available for human manipulation. With his success, people became their bodies, inanimate objects composed of dead matter. That was the reality he had fathered.

The night that his creation had opened its watery eye and breathed its first and Frankenstein had fled his lab in horror, he fell into a wild and fitful sleep. In his dreams, he thought he saw his beloved Elizabeth in the bloom of health, walking the streets of the town where his lab was. Delighted and surprised because she was in reality many miles away, he embraced her, but with the first kiss, her lips became livid with the hue of death; her features changed, and she became his long-dead mother. A shroud covered her, and he could see the grave worms crawling in the folds of her dress.[14] The monster killed everyone that Frankenstein loved twice; the first death, which occurred when the monster opened his eyes, was much more terrible than the last, which came directly at the monster's hands. If Frankenstein could create human life out of dead matter, equating the two in his chemical equations, then, by simply reversing his equations, everyone that he loved was dead matter and object of human command—something to be used, not loved, and nothing more.

As the story develops and the monster kills everyone that Frankenstein loves, he only completes what Frankenstein had in truth already accomplished with his creation. Throughout the rest of the story, with each succeeding death of a beloved friend or family member, Frankenstein is overcome with guilt, and he says repeatedly that he has killed them.[15] Everyone dismisses this as incoherence inspired by grief, but Frankenstein is entirely correct and entirely coherent. His creation, by the sheer truth of its being, has killed everyone he loves by destroying their humanity, turning them into an object for Man's utility. The monster is Frankenstein's double, the object of his subjectivity, and the re-presentation of his power as technician. On the surface the double appears as a negation of the self, but more profoundly, the double is a completion of the self, an other, which combined with the self creates a whole. Each exists only because the other exists. Dissolving this difference, making subject and object an identity, the monster is monstrous for Frankenstein because it reveals everything as the will's utility. The differences that made his quest for knowledge matter disappear.

Despite its gentleness, its care and concern for humanity, its overwhelmingly human need for companionship, Frankenstein's creation was doomed to become a monster, a terrorist at war with society, because of its very Being as a Man-made fabrication. In its truth as inanimate matter made living by the hand of man, the perfect representation of humanity, it was the incontrovertible denouement of Man as subject and archytect of the world. As the highest accomplishment of Man's technology, the fullest demonstration of his subjectivity, the monster was the other truth of Man, the representation of Man's being as the object of his own technology. Throughout the book, Frankenstein seeks only to accomplish the full humanity of Man, to protect human subjectivity.

When he realized that his creation was not the instrument of humanity's freedom but the demonstration of its slavery, he sought to protect humanity from the terrible truth he so brilliantly demonstrated. This is what Frankenstein was doing when he refused to give the monster the female companion that it desired. Creating a companion to the monster, so that he could father his own children, would only be a further demonstration of Man's nonhumanity.[16] Then, it could perfectly duplicate and represent the reproduction of the human species. In order to prevent this perfect Man-made re-presentation of the human condition from taking place, Frankenstein had to place a limit on his power to manipulate in order to protect humanity from the subjectivity of Man. Risking his own life against the rage of the monster, Frankenstein breaks his promise and rips apart the almost-completed female, seeking, as always, the greatest benefit to society rather than his own. Man must be master, and nature must be represented as his object. Any confusion of this essential difference, as the monster was, must be suppressed, denied its re-presentation.[17] Otherwise the truth of what Technoarchy was actually doing might be revealed.

This is the final insult to the monster. Knowing that he will never be anything except his utility, he begins his reign of bloody terror and revenge. Contrary to the myth, the monster was not created evil, his brain coming from a sadistic criminal, but he became that way because of his relation to technical Man as his other. Frankenstein's creation developed as Rousseau's natural Man developed. At first he was benevolent and good, rescuing a child from a raging torrent and secretly helping an impoverished family with its housework, but as his social consciousness developed and with it his awareness of the cruelties inflicted on him by society, and especially his creator, who made him only as a means for his knowledge, he was possessed by bitter resentment.

Everywhere he saw bliss, from which he alone, the hideous other, the object of Man's utility, was irrevocably excluded. At first benevolent and good because he wanted to participate in society, to have friends and to be a friend, his exclusion and his Being as utility made him miserable and a fiend. In contrast to Rousseau, however, the monster became evil not because society made him that way but because he was excluded from the humanity of society because he was its other, the demonstration of its subjectivity. He became a terrorist bent on revenge because that was the identity forced upon him.

For Rousseau, an author that Mary Shelley is obviously deeply indebted to, people are by nature isolated and solitary, yet as they mature and begin to develop their social nature, they cease to be alone. In fact, in the fuzzy transition from the state of nature to civilization, in which the dividing line between the natural and the artificial and social is never clearly drawn, it is the social passions, naturally derived, which lead people to form the social contract.[18]

Though the legitimacy of the social contract rests upon the interests of individuals, its formation rests upon the recognition and humanity of others, upon the moral claims that others have on the self's actions, beliefs, and commitments. For Rousseau (and Mary Shelley, I think), it is love and hate, the fear of death and the ambition to dominate—all socially made emotions, not given by nature—that lead to the development of language, the formation of the social contract, and the evolution of society from equality to inequality. It is not pure self-interest that puts society together, nor individual calculations of marginal utility, but the passions, progressively developed within society, that make the ties that bind, the reason that considers, and the morality that obligates.

For Rousseau and Mary Shelley, as indeed for the entire epoch of Technoarchy in one way or another, our moral commitments to one another, embodied in social conventions, are the products of society, and within the notions of property, love, hate, ambition, and friendship there is the implication of intersubjectivity, community, and the recognition of another as a will deserving respect.[19] Although Mary Shelley's monster is a social outcast, he nonetheless develops into a human being, one fully as human as anyone else. This, given what society must insist upon, is his great tragedy. He belongs to it, but does not belong. He has all the emotions of a social being, but because of what he is, he cannot be permitted the chance to express them.

Though human, his Being as human creation irrevocably excludes him from humanity, community, and love. To keep safe the dichotomies Man's reason requires to dominate its world and free Man for his mastery, the monster must remain in his place as humanity's object. As his self-consciousness grew, the monster became aware of his differences and of the act of exclusion that denied him his humanity. An artificial creation, attached to no place, no family, no home, the monster had no relationship to anything, except his experimental utility to his creator. He had no caring father to watch over his childhood, no mother to bless him with smiles and caresses, no one to lament his pain or his annihilation.[20] The first Man produced by the new technology was, despite the sociality of his passions, simply an atom, an island of sheer existence surrounded by an endless sea of his utility. As he increasingly became what he was, a human being, he increasingly became a monster, a terrorist at war with his humanity.

As the monster became aware of what he was and how his being as Man-made man made him different, how it excluded him from his humanity, his awareness increasingly became a torment to him. What was he, he wondered? A monster certainly; something different, something horrible and evil—a fallen angel, perhaps? No, not even a fallen angel. Even Satan had his companions, fellow devils to admire and encourage him. Cut off from society, excluded from any relation that could satisfy his social passions, knowing himself as a monster, despair and desolation took the place of his kindness and benevolence, hatred and rage the place of his love and sympathy. A howling revenge, borne by the hell within him, became the purpose of his being. Unable to give love or get it, he could still give pain and death and get terror as his reward. Made human but denied his humanity by his role as other, the monster declared everlasting war against the humanity of Man, and especially against the humanity of his creator. Radically repudiating every value and meaning that humanity had celebrated and in which he had joyfully participated, the monster became the active agent of nihilism. If he could not be as he wanted, a human being with a home and a place, no one else would be either. Everything human would be destroyed.[21]

With no-thing thinging for him, no gathering of care to restrain him, no home or god to give him a responsibility for anything, but knowing everything that he lacked and haunted by it, the monster began killing with a devilish despair.[22] He did not strike directly at Frankenstein, killing him cleanly and directly. That would be too easy and kind, too human. Instead, he struck at the love that surrounded Frankenstein, the love that reflected his humanity and gave him a home and a place. And as a result of each murder, Frankenstein became more like the monster, a homeless, loveless man, possessed by a raging necessity for revenge. A slave to his hatred and to the instrumentality his creation had imposed on him, Frankenstein became the monster he despised. Just as Frankenstein used his knowledge of nature to master lifeless matter and create a man, so the monster, now a subtle scientist of morals, used his knowledge of society to manipulate Frankenstein, to reduce him to an atom of revengeful utility, an other whose pain was testament to his mastery. And thus the master becomes slave through technical knowledge, and the slave becomes master, round and round in an inward spiral of technical mastery that denies everyone subjectivity and eventually closes in on a despairing death for all in a cold Arctic wasteland far from any human habitation.

Such is Mary Shelley's warning and prophecy. Technoarchy fathers monsters by reducing humanity to slavery. Unlike other possible ways of being, the doubles that it fathers are truly monstrous, hateful, rancorous beings bent on terrible revenge because the humanity that this age exalts is denied them, since their only Being is as utility. There is something seriously wrong with the utopia that patriarchy's science and technology would build.

Perhaps Sade would have understood Mary Shelley's monster and its need for revenge. For Sade, the pain and horror of other people, inflicted by a master of torture, is a delightful intoxicant, a sign of difference, hierarchy, and inferiority that separates them from the godly self. Other people's pain, degradation, enslavement, and death are testaments to their utility, a subtle demonstration that they have been mastered. Power is the truth of the other's pain, and rape, whether actual or symbolic, is always its technique.[23] The problem Sade, as the literary monster of our time, poses for Technoarchy's utopias of mastery is this: how can all the economies of production, hierarchies of control, and systems of order, which hold the individual as the Reserved and that curtail, constitute, and circumscribe her actions as its necessity, justify its morality?

If there is no truth to morality but its own contingency, if the only meaning that exists for humanity is what it gives itself as convention, if morality is Man-made and fathered by our history, then what is there in Man's humanity that requires us to submit to it? If morality is a tool of human control, as it is for a scientist of morals like Rousseau, a tool that must be made subject to the will of Man, why must we submit to it in a herd? Why not assert our mastery as an individual and make our own morals as our nature prompts us? What is it in modern morality that requires us to recognize the humanity of others, to accept moral virtues, to subordinate our private will to the general will? The truth that, being members of society, we are inescapably implicated in our humanity, that we are moral beings and have always been so, says nothing about why we, as the masters of our morality, should continue to be such. Our mastery of our morality, reflected in our Being toward it as social convention, entitles us to make of it what we want. Our technical Being toward morality, our recognition of it as utility and tool, gives us the license to do as we, as individuals, will—whatever we will, whether the tortured death of the innocent or the genocidal death of the multitudes.[24]

Because it knows everything as human instrument, Technoarchy has no argument, except its own power to censor, control, and render the other monstrous, directly to counter this challenge to it with, since it itself is the will to mastery through human technology and it itself makes everything into utility. Rousseau, Sade's most venomous double, formulates in the social contract the conditions in which Man can be master and moral at the same time.[25] Indeed, he goes so far as to make Man's mastery contingent upon his social morality. For Rousseau, like Hobbes, Man's moral obligation to others is derived not from immediate nature and not from God but from the agreement that makes society possible, the social contract, which is the most secure ground of the will. Rousseau, like Sade, seeks only to establish the reign of Man's mastery.

Contrary to Sade, the natural fact of physical power or force does not in any way legitimate or justify the rule of the strongest. Although force is a physical power, as natural as it is real, no morality can result from its effects because yielding to it is an act of necessity and possibly of prudence, not of human will. And although Rousseau is willing to agree that all power comes from God, he is not willing to agree that that derivation makes it legitimate, because God is as much the source of illegitimate power as he is the source of all illness.[26] Just as we need not hesitate to call a doctor when we are sick, we need not hesitate to build a legitimate state.[27]

Much as Nietzsche did later, Rousseau locates Man's freedom in the correspondence of his will to his Being as master of himself, not in a supernatural beyond.[28] Because Man's Being is social, the circumstances of his freedom and his willing must be social. Since force produces no right, nature no natural authority, and God no special privilege, Man must find in himself his own source of legitimate authority. And the measure of his legitimate authority will be the extent to which his state establishes and makes possible Man's correspondence with his being, his freedom as master of himself. Man realizes his place as his will willing itself when the general will, the commitment to the common good, is identical with each particular will.[29]

This occurs only under the most ideal circumstances—a small, egalitarian, self-sufficient, isolated, agrarian state that has had a long history of conventions and traditions that support democratic participation in the affairs of state.[30] Under these conditions the social contract, subordinating the ends of each to the goals of all, defends and protects the person and goods of each associate, even while each individual obeys only himself. The act of association, in which all agree to subordinate their private will to the common good, produces the virtues and the moral obligation necessary to fulfill the terms of the social contract. The general will, through the developmental act of association, is the means to freedom for each individual. Because their freedom and humanity as human beings are realizable only through their sociality, individuals can be, and should be, forced to be free if their particular will differs significantly from the general will.[31]

Children of Technoarchy, both Rousseau and Sade recognize its first truth, the Man-made nature of humanity—the artificial nature of his morals and the necessity of grounding everything on a technology of the will.[32] For both Rousseau and Sade, Man is the master of himself and, through himself, of all the earth. This certainty that the will is the fundamental reality of Man is the truth that unites them throughout their most extreme differences. While Rousseau recognizes the necessity of placing limits, an archytecture of its own possibility, on the individual will in order for it to come to its own truth, Sade rejects any limit to the will's willing—not God's commandments; not Man's own Being in his conventions, morals, and traditions; not even nature's promptings. Since everything is human instrument, everything is to be as instruments are used, for the will's willing. Since the truth of humanity is the degradation and instrumental use of everything human, Man is entirely within his right to use the humanity of others as the satisfaction of the monstrous passions his own Being as utility for others has brought forth in him. Sade is thus the ultimate truth of Technoarchy, the pure expression of it as the monster whose willing knows no limit. Because it knows no human limit, it cannot be universalized in a general will or archytecture of any other sort but must find its satisfaction in the unlimited nihilism of a unique will, in the unbounded archytecture of a single governing will that knows the entire world as its own instrument of play. The unique one knows itself as master, and giddily affirms itself as such, by its violation of everything that opposes its willing—Man's virtues,

God's will, nature's inevitable death, and the sanctity of innocence.[33]

The central point around which Sade's Being revolves, the main archytecture of the modern self that he celebrates, and yet secretly despises, is that he is alone in his willing—profoundly, awesomely alone. So alone that not the most terrible agony or need of other people can penetrate the isolation of his will, the archytecture that he has surrounded himself with. Between the will of the unique one and all others, there is an unbridgeable chasm, a separation of the experience of self that transforms all others into the will's utility.[34] The most terrible agony of others, unexperienced by the unique one, is nothing to it, while the faintest touch of pleasure that is felt is everything and should be preferred to the universal sum of others' miseries.

Seeking Technoarchy's freedom through the archytecture of the will's willing, Sade has a very complicated relationship to other people, much like the master in Hegel's master-slave dialectic.[35] On the one hand, their will cannot be known as a will, since they are inescapably other, the object of the will's utility; on the other hand, their will can be known because their pain, experienced as *their* pain, is a demonstration of their otherness, their objectivity, and of the unique one's mastery over them. The consciousness of the other is a necessary moment in the master's affirmation of his will and freedom.

In his zest to dispossess other people of their will and freedom, to prove that they are nothing but the will's utility, Sade betrays his profound dependence on other people, if only because he needs them as victims to feel his power, to acknowledge his mastery in their slavery. If other people actually are nothing but their utility, why is pleasure gained from torturing them? If other people are just as noble or ignoble as the worms that chew on their corpses, why won't the worms do as well for a ventilation of Sade's energy? No doubt it is because other people are singularly capable of experiencing and knowing the power of the Master. In its final certainty of itself as master, as the command that commands, the will can know itself as will only through the destruction of another will.

This is the depravity Technoarchy comes to when it finally comes to know freedom as only human control. As objects of utility, other people are especially valuable because their consciousness knows their degradation and subjugation and through it acknowledges the master's mastery. Worms do not. People can know themselves as slaves, tools subjected to necessity and mastery. Worms cannot. The unique one's mastery and freedom depends upon the denial of it to others, even though it is also dependent on the recognition of it by others.

Although others are nothing, Sade writes books; he seeks the recognition of others, desires to be their desire. The vague realization of this torments him, driving him to the flash of an anger and the depth of a hatred that is truly monstrous. Sade needs Rousseau because his humanity is the perfect object and tool for Sade's master. The virtue that Rousseau celebrates is useful because it can be violated, demonstrating and affirming the mastery of the unique one. The laws which Rousseau saw as being the means by which humanity could come to its freedom, being the truth of its implication in the general will, are for Sade a constraint on his freedom, denying him his will just as they make possible the full expression of his will. Forged for universal application, leveling all differences to the tyranny of passionless reason, laws are by nature in perpetual conflict with the individual's will. Contrary to Rousseau, the laws the general will fabricates for the individual's freedom are in truth an unnatural tyranny over the will. Their being is entirely alien to the will's Being as its own master.[36]

If on occasion the laws protect the will, they more often hinder it, trouble it, and fetter it, denying it the limitless assertion of its mastery. For Sade, the archytecture of Man-made laws are secondary to the archytecture of the modern self, of Man fathering himself. Sade goes to such a length in reversing Rousseau's notion of the law being the will's means to come to its freedom as master that he claims it is wrong for the law to kill, while it is acceptable for an individual to do it, for the law is alien to the will and opposed to it, while the individual's natural passions, which inform their act of murder, are not.[37]

The particular will, being closer to its truth as master, is always the true means to human freedom. Consequently, any limit placed on it is a tyrannical limit, unjustified and false, placed on human freedom.

For all his moral nihilism, Sade is not that much different from Rousseau, or any other figure in the Enlightenment. He only has the perverse and ugly courage to think what the will willing itself really means—the emptiness, the violence, the pain. If Rousseau and Hobbes are utopian, dreaming of places of legitimacy that never existed and can never be, Sade, who is no less utopian than they, insisting as he does on the absolute power of the will to will itself, reveals the monstrous absurdity of Technoarchy's quest for a Man-made utopia, with its cruelty, violence, and utter lack of care and friendship. These are part of the archytecture of Man, unavoidably so—only Technoarchy represses and conceals this, throwing it into its shadow. But it is present in the factory and all the other disciplinary institutions it builds, places where people are treated like machines, simply a means for the will to will itself. And the people in these places do suffer terribly. Their well-being, health, dignity, families, and futures are all sacrificed to their utility as instruments of power. That is the violence that Sade reveals.

We would be fooling ourselves if we did not admit that all too often people in positions of power *like* abusing those they have power over, *like* making them grovel. Hurting them, humiliating them, and taking away their jobs, dignity, and self-respect make them feel powerful, in control, free to do what they will. Unavoidably having been at some time a means to someone else's will themselves, they get revenge for it by giving their pain to others in turn. And the nightmare goes on, each victim passing his or her pain on to the next. Sade's monstrous cruelty is very close to us. It is present in unemployment lines, deadening work, spouse abuse, child abuse, sexual harassment, every assortment of cruelty from rape to murder, and in the destruction of the earth itself. None of us in this age, I fear, has escaped the sadism of someone who has gloried over our helplessness, and few of us can honestly claim to be completely free of the sadist within, to refuse to glory in pain inflicted on others—that is the ugly shadow of freedom as control. Until we acknowledge this shadow in our life, accept it as part of ourselves and our thought and act to heal it, none of us ever will. We must become friends to each other, rather than each other's tool.

CHAPTER ELEVEN

The Turning

Dreams and despair
swallowed by rotting tongues
and stay silent
Forbidden monsters
the endless days,
numb routines,
and dying lives we live
And then the world turns
and hope stalks its prey
on kitten's paws,
clumsy and playful,
then powerful and strong
And dreams
become the most possible
things in the world

—the author

CCORDING to the Thinker, it is not our truth as a metaphysics of command and control that shall set us free, but rather it is this truth, this insistent utopia of universality and eternity, that enslaves us, imprisoning us with its necessity. The idea, so obvious that we cannot deny it, that truth is the means to freedom, the way to power over all the earth, is for us in our time a prison. By making the world into something to be controlled, we also make ourselves into something to be controlled, an object of discourse and discipline. And our souls become, as Foucault observes, the prisons of our bodies. We shall not escape the terrible fates that may await us—nuclear war, ecological disaster, economic collapse, political totalitarianism—by means of this truth of the world; we only assure something like them with it. Indeed, unless we turn from our way, we will die of this will to truth. As a result, we must think about the question of political action, of how we are to live this turning. Other truths, other ways of letting the world world, need to be brought forth. In particular, we need to think of the ways in which freedom is friendship and letting the world world, not will and control.

There is something strange (and something hauntingly familiar) about the debate over the Thinker's Nazi past. Informed by a passion that suggests that something considerably more than a correct reading of the past is at stake, it is raising (and obscuring) questions about the relationship between politics and philosophy, the link between thought and virtue, and the essential character of those who would betray reason. Around this debate, haunting its margins, yet penetrating its core, there are protests of violated innocence, vindictive demands for purity, and angry judgments against evil. Somebody is clearly having problems with their shadow. And perhaps the people making the accusations are telling us more about themselves than they are revealing about the Thinker's involvement in Nazism.

First, why is it that of all the intellectuals who got involved with the Nazis, only the Thinker is signaled out for examination? Prominent positivists, analytic philosophers, neo-Kantians, neo-Hegelians, to mention but a few, all used their concepts to elaborate and justify Nazi politics.[1] And yet, only those schools of thought that are influenced by the Thinker are made to respond to the question: are they inherently fascist, somehow contaminated with the Thinker's sins?

Second, everyone admits (to varying degrees) that Victor Farias's book *Heidegger and Nazism* (which seems to be the focus of much of this debate) is a poor reading of the Thinker's work. No one recommends it as a model of scholarship, not even his editors in their foreword to it.[2] As critic after critic has pointed out, there are endless distortions, exaggerations, and seedy attempts to discredit the Thinker merely by associating him with the people around him.[3] If the debate was not really about something else, if it was only based on the new "facts" it reveals, the debate surrounding this book would, I suspect, be nothing more than a tempest in a toilet bowl. And yet, everyone (even Farias's critics, by means of their vehement opposition to it) agrees that the book is a monument, an event that, along with the De Man controversy, marks the end of innocence for postmodernism. Now, it is implicated in, linked with, and contaminated by politics—the most dangerous kind of politics too. Of all things, and of all the unthinkable ironies, an exclusionary politics seeking to abolish difference and otherness by means of hatred, fear, and naked power. Farias's book has become something to endorse or oppose, something that signifies one's place in the debate over our future. Clearly, Farias's book is a book to be used, not read, certainly not carefully or thoughtfully. The question is, What is it to be used *for*?

Identifying, as many people in this debate are at least implicitly doing, postmodernism with Nazism—with totalitarianism, torture, genocide, and atavistic aggression—is a dangerous thing, whether they support this identity or oppose it. Regardless of the reader, regardless of postmodernism's doubts about the authority of the author, regardless of the questions that can be raised about the singularity of any unity or text, this debate frames a context for the reading of the Thinker's work, forcefully summoning his work as a whole under its interpretation and judging it under the exclusive terms of his politics and his failings as a thinker. It links, then poisons, everything with all the horror, fear, anger, and resentment that the abominations of Nazism have come to signify, making it into a monstrous abomination. In the end, it makes quiet and meditative thinking difficult, if not impossible, and destroys any serious encounter with the dangers of modernity that the Thinker tried to reveal. It leaves us lost in our shadow, thoughtless, friendless, unable to let truth happen.

Perhaps this failure to think, this unwillingness to let the shadow be, after all, is the truth of what evil is—opposition to one's hated other, whatever it is. Originally, as the dictionary tells us,[4] the meaning of "satan" and "devil" were not as pejorative as they are today. Their root words in Greek meant only "opposition." "Satan" meant an adversary. To oppose something, by the mere act of opposing it, was to be evil, to bring evil forth.[5] The content of the opposition was irrelevant because the good and evil that it assumed was itself constituted by the act of opposition and the hatred from which it originated.[6] As it is said in the "Gospel of Philip": "Light and Darkness, life and death, right and left, are brothers of one another. They are inseparable. Because of this, the 'good' are not good, nor the 'evil' evil, nor is 'life' life, nor is 'death' death."[7] Good and evil are not differences as much as they are identities. Invoking one invokes the other, giving it form, purpose, solidity.

If we accept this, it is not such a strange and twisted irony that the greatest events of evil—and truly the Holocaust is one of them—are always fathered by those seeking to identify, isolate, and destroy whatever they believe is evil, to make the world pure and good, obedient to its true origins. Certainly that is what the Nazis wanted, as did the Spanish Inquisitors, the Crusaders, the witch-hunters, and our more contemporary cold warriors. There is something tragic and inevitable about opposition to evil that draws evil to itself, setting it up to play its role of angry, resentful, and vindictive destruction. Whenever there is something to be opposed, evil is the first to judge it evil, the first to hate it, and the first to take the lead against it. And the more absolute the power of the evil opposed, the more evil and powerful the opponent to evil becomes, drawing its own strength from its opponent.[8] Evil is simply the shadow of good, what it is not, what it represses and must not admit of itself.

Perhaps this hatred and fear is the nature of all opposition. Perhaps this is why thoughtful, caring, self-reflecting people so rarely engage in it. It is too corrupting, too alien to their nature. But what are we to do when the shadow is all around us, menacing and horrible, when there is an unrelenting evil to be opposed, a vast evil that is destroying everything holy, noble, and good? What are we to do in a world endangered by hunger, the bomb, and environmental catastrophe, by totalitarianism, torture, and obsessive thought control? The great question is how to oppose evil without becoming evil, how to oppose hatred without hating, to oppose totalitarianism without becoming totalitarian, to save the earth without destroying it? How are we, putting it into the Thinker's language, to keep safe the thinging of the thing amid the darkening of the world, the flight of the gods, the destruction of the earth, and the transformation of humankind into a mass?

As I think about the Thinker's life, and about the undeniable evil of his political involvement with the Nazis and of the use that is being made of it now, I think about how, perhaps, I should not mention this identity that is linking postmodernism with Nazism, even to oppose it. I think about how I should not claim my rights as possible victim of it to speak out against it. Perhaps, by opposing it, I will only make the venom that sustains it stronger, more poisonous—and risk my soul in the process. This appears to be Derrida's position.[9] Much too much has been written about Farias's book. Those who have opposed Farias (and Derrida knows that he too is at least as guilty as anyone of this—no one can let go of it) are responsible for making his work into a monument, a center of debate that swallows everything up, makes everything ugly.[10] In his book *Of Spirit*, Derrida exiles Farias's name, putting it under erasure, but his spirit, an unfriendly one, still haunts every page. His presence is made conspicuous by his absence in a text that makes much of this strategy, refers to it continually. "I'm thinking," Derrida writes, "in particular of all those modalities of 'avoiding' which come down to saying without saying, writing without writing, using words without using them."[11] I have no doubt that Derrida's strategy of erasure is a good one—maybe it will even work eventually. But I, risking the shadow, am going to pursue a more direct strategy, follow a more dangerous path, praying that "as the danger grows, so grows the saving power."

Let me, then (and please do forgive me my sins against

myself), say what I should not, oppose for a moment an opposition that I want forgotten, an identity that I want undone. That this identity—the Thinker and Nazism—is asserted so aggressively by so many people seems strange, as if some other strategy were going on than the simple disclosure of truth, as if it were more an attempt to silence, marginalize, and exclude certain thoughts than anything else. How can a body of thought that made way for the deconstruction of totalities possibly be totalitarian? How can a thinking that claims that the essence of truth is its other, untruth, be implicated in a politics that would father death camps to eliminate whole categories of people? The Thinker's thought, especially his later thought, does not appear to be a likely candidate to father a world of exclusion or sustain a politics of fear, hatred, and resentment. On the contrary, it itself presents itself as a strategy for saving us from the dangers of modernity, the totalitarian need for exclusion, control, will, and purity that are themselves fathered by the logic of all our scientific and technical systems of reason.[12] It resists the fear that runs from alterity; it is open to its other and in fact seeks it out, celebrates it, depends upon it, is grateful to it. And it is careful to spare and protect it. It is not afraid for its fatherhood, does not need to claim it, protect its origins with laws of descent, or insist upon hierarchies to establish the legitimacy of its reign.

Technoarchy, more or less like every metaphysical tradition or archytecture that has preceded it, is as totalitarian as it is exclusionary and purifying. Ambiguity, mystery, slack, chaos, and equivocality are all its others that it seeks to dominate, marginalize, and eliminate. According to the Thinker, it divides the world up into that which is its own truth and that which is not, and then it sets out to make the world pure, to master it with a "true" expression of its relentless and universal logic.[13] Uncompromising opponent to everything it is not, reason must oppose itself to its irrationality and never fail to overcome it.

The Thinker's life reflects two ways of turning: as demanding submission to spirit, and as a letting be of spirit. The first strategy made him into a political activist, a monster that supported and legitimated the Nazi state; the second, into a quiet meditative thinker who thought a way that let the world world and yet remained a critic of the archytecture of modernity. In his "Rectorship Address" in 1933, the Thinker spoke of a decision in which the Germans, as a historical-spiritual people, gather themselves up and will their truth.[14] Every individual participated in this decision, even those who evaded it. On this decision the entire fate of Western civilization, he thought then, hung in the balance.[15] This decision to bring themselves into accord with their truth had to be made by the German people. Not revealed clearly by Being, not known or properly understood by all, it was a choice that they had to be led in making. So the principle of leadership displaced "academic freedom" and made it obligatory for students and "the little people" to obey those more in tune to Being.

As Arnold Davidson argues, the Thinker breaks radically from his earlier understanding of the self in *Being and Time* from what he asserts in the "Rectorship Address."[16] In *Being and Time*, the Thinker argues extensively that there is a tension between the authentic self and the self that is lost in the apparent thoughts of others, the way of being the Thinker calls the they world. Through resoluteness, the self recovers itself, returning to its "ownmost potentiality-for-Being." This resoluteness, this authentic moment, according to Davidson, is completely set aside in the "Rectorship Address." In its place there is an unequivocal command to submit to authority, to obey the leader, and to ignore any inner stirrings of personal conscience. Anxious to protect the claim of the father, the Thinker insists that what is needed is the "genuine following of those who are of a new mind." Without this distinction between authenticity and the they world to protect it, the individual human being, the "I myself," is given no space for independent thinking, and its thought is usurped by the people, the state, and the German fate. Everything must be brought into accord with the Fatherland, as it is represented by the (enlightened) leader. *This* devotion to the Fatherland is what made Heidegger into a Nazi—for a while.

But only for a while. His work after 1945 can be read as an

attempt to reappropriate his vocation as a thinker, responding not to the claims of the Fatherland but to the gentle whisper of the world's worlding. The thinker now responds not to the existence of the state, the people, or the German spirit but to Being itself, in all its anarchy and dispersion. Nationalism, along with internationalism, individualism, and collectivism, becomes something to overcome, an archytecture to think beyond. The profound and radical homelessness of the modern age becomes the problem he addresses, not the decisions facing any one nation, state, or people. And never again does the Thinker appeal to the authority of the leader to lead; rather, all appeals now come from Being itself, not any person or archytecture. The thinker is thus freed from any of the demands of nation, state, people, or authority.[17]

But freedom from such demands is only part of the movement toward true freedom. For thinking to come to true freedom it must call into question modern reason and the entire metaphysical tradition, since it has made possible the oblivion of Being found in the Soviet Union, Nazi Germany, and the United States—the three great articulations of modern ideology. As the Thinker argues in "The Turning," a pivotal essay written after he had completed his turn, it is not modernity's reason, with its awesome power for total command and control, its relentless need to displace its others with its placeless and timeless truths, that shall set us free and save us from the dangers and evils of our age. Rather, our reason itself is our danger, the origin and domain of the evil that traps us in the oppositions that build, and make logical, our factories of death, whether they be death camps, nuclear bomb factories, or our toxic agriculture.[18] Having built all the things our age knows, cannot deny, or imagine otherwise, having set up a whole world of energy, communications, transportation, and food systems, our reason has turned on us, possessed us with its purifying imperatives, made us dependent on its unequivocal power, and imprisoned us in its clear and crystalline logic, enslaving us to its totalitarian truth, necessity, and doom.

The modern idea—so obvious and rational that we cannot deny it without seeming to blaspheme all the progress of our civilization—that submission to reason is the means to freedom, the way to safety from any danger, is now for the Thinker a monstrous delusion and an entrapping prison for thought.[19] By means of reason, we shall not escape the terrible evils that reason itself knows may be waiting for us—nuclear war, ecological disaster, economic collapse, political totalitarianism—and especially not the biggest threat of all, nihilism. Thinking of these things rationally, we only assure them yet again, yet more surely. Indeed, unless we turn from the way of reason, we may all die of the world of oppositions it has imprisoned us in.

The power that our reason gives us is a subtle poison that corrupts our Being, opposing us to the earth, our truth, and our place. Making secure the opposition between reason and its betrayal, Man and chaos, good and evil, it cuts us off from that which can save us, the integration of our shadow. And the most dreadful, dangerous, and monstrous truth of modern reason is that it does not reveal itself as a monstrous doom.[20] On the contrary, it seems evident to all that modern reason, and the technology it spawns and makes identical with itself, is but a simple tool, a value-neutral means to power in the hands of Man, and that, properly applied, it can save us from any evil, any opponent.[21] We only have to use it properly for it to save us. But Man, as the Thinker (and through him, Derrida and Foucault) has taught us, is not the master of his fate or his technology; for it is Man himself, as new lord and master of all the earth, that is ordered forth by modernity, a way of Being that precedes and makes possible Man's mastery.[22] For the Thinker, at this time in his thinking, there can be no salvation that comes from submitting to a leader's authority. Such submission could only be a yet purer expression of human imprisonment in the quest for mastery.

As master of the earth, the positer of the universal and eternal categories by which his whole world becomes available purely as his utility, Man is the subject of a subjugation more profound than himself.[23] He is not the master of his technology, the will that orders its use, but, as master, is the one mastered, enslaved, and ordered into use by modernity, enslaved by the logic of reason's opponents.[24] Before Man can possess the world with his reason, he himself is possessed by reason, delivered over to its metaphysical organization of truth, its logic, its power to designate, name, and create the differences, hierarchies, and oppositions that dominate the ordering of things. The concealed monstrousness of this quest to master the whole world led the Thinker to write: "Agriculture is now a mechanized food industry. As for its truth, it is the same thing as the manufacture of corpses in the gas chambers and the death camps, the same thing as the blockades and reduction of countries to famine, the same thing as the manufacture of hydrogen bombs."[25]

Lacoue-Labarthe, in his book *Heidegger, Art, and Politics*, misunderstands the full horror of what this is saying. According to Lacoue-Labarthe, there is an incommensurable difference between the extermination of Jews in death camps and the way we farm, or even the way we build hydrogen bombs. The horror of it is entirely different, not at all technical, but aimlessly nihilistic. In his reading of this quotation, the Thinker is failing to acknowledge the awesome pain and suffering and tortured deaths that millions of Jews underwent quite aside from any rational or technical purpose. The annihilation of the Jews, an entirely heterogeneous population, without any sort of systematic linkage or purpose, was completely irrational, based on nothing but hatred and fear. This, according to Lacoue-Labarthe, is the true horror of the death camps—they served no purpose at all.

Perhaps the Thinker is slighting the horror of it, possibly to mitigate his own implication in it. Counter to Lacoue-Labarthe, however, the extermination of the Jews was not purposeless but a necessity of a world that reason has built. Denying its shadow, reason must fear its others, hate them, try to destroy them. It feels persecuted by them. And if Jews were the main targets of the Nazis, the archytecture of their hatred was pretty much the same for Communists, Gypsies, homosexuals, and so on. (Contrary to Lacoue-Labarthe, the purpose of the death camps was not overwhelmingly to kill Jews. Only about half of those killed in them were Jews.) The slaughter of all these people was not purposeless; the Nazis knew exactly what they were doing and could give lots of *reasons* for it. It all really was very logical. These Jews, Gypsies, Poles, homosexuals, and so on were contaminating the purity of the Aryan race; they had to be destroyed.

The otherness of these people, what made them targets for genocide, was constituted and structured by the archytecture of reason, which is totalitarian, exclusionary, and extinguishing. Fathered by the fear that reason has of its others, the Holocaust could happen only in the modern age the way it did. Jews have long been hated in the West and have long suffered tragedies like the Spanish Inquisition, but the dream of extinguishing all of them is an extreme possibility made necessary only by modernity's unbounded need for purity. Nor were the death camps mutations unique to the Nazis. In their organization and their need for isolating and extinguishing otherness, they are almost indistinguishable in structure from the internment camps that the U.S. government used to isolate Japanese-Americans during the same war.[26]

Again to the point that Lacoue-Labarthe makes about the Holocaust being unique in its quest for purity,[27] is there really any essential difference between the Holocaust and our preparations for a nuclear holocaust, except the accident or crisis that would start it? The Jews are dead, and the billions—very possibly all of humanity—that would die in a nuclear holocaust are not yet dead, surely, but what *essential* difference is there between past attempts to purify the earth totally of otherness and future attempts, which a nuclear war would surely do? Once we imagine what the future may bring because of what the truth of our age has made possible, Lacoue-Labarthe's incommensurable difference becomes nothing more than a known body count versus an unknown one, nothing more than the horror of a known past versus the horror of an unknown, but possible, future. Thought essentially, there is nothing terribly unique about the Holocaust. And the future effects of our mechanized agriculture, which must be so negligent of the earth, so careless of the life-giving soil it exploits, could easily surpass the horrors of the death camps.[28]

But the fact that we are imprisoned in the world reason has built does not mean, according to the Thinker, that humanity is forever delivered helplessly over to commanding reason, pure slave to a master that does not die.[29] On the contrary, we are doomed to reason's slavery only as long as we are called to be its masters. Our unique danger is that our way of being, Man and his reason, does not allow us to know that our understanding of freedom as rational control, as opposition to that which would escape control, is the destiny trapping us in the evils we would abolish. Modernity's reason possesses us before we can master it and denies us the possibility of thinking any thought that says that our freedom is not our salvation but our prison. Knowing freedom only as Man's command and control, we thoughtlessly push on with it, reasoning that somehow, some way, it will save us. According to the Thinker, this danger, the domain of reason's oppositions, has been our destiny for so long that it is in its own way moving toward a decision that will reveal another destiny, enabling us to think of freedom in another way than as Man's control, as opposition to that which thwarts reason's reign over the earth.

A destiny blooms forth in its own way from the earth, and the world it becomes is continually adapting itself to it. As a world, a destiny carries on a dialogue with the earth, revealing it until the earth breaks its limits, its time, and becomes another destiny, another world. Destinies do not change in a rational way, limited by the logic of oppositions they must assume in their own time or at the hand of a powerful leader, but they break radically and incommensurably with each other, becoming a destiny that cannot be any other destiny than its own.[30]

But if a change in the destiny of modernity occurs and another destiny, another way of Being that does not reestablish reason's enemies, breaks forth and takes its place, this, according to the Thinker, does not mean that the technology whose truth for us lies in modernity will be done completely away with, that we will necessarily assume the most primitive way of life excluding any use of things, abandoning all the power reason has given us over things.[31] In its truth as the way of revealing truth, and not merely as it is revealed by modern reason, technology is the way of human Being on the earth; humanity cannot be without bringing things forth from the earth.

But we can think about our technology and we can adopt a way of bringing things forth that spares the earth, preserves the world, and leaves the mystery untrapped by the dangers and oppositions of modern reason's logic. Thought carefully, the technology our reason deploys is neither the means by which we attain our mastery nor the means by which we are mastered. Always preceding it, it is something wholly other than the dialectic of master and slave, of opposition and control, and especially of leader and follower, because it is something other than the Man-made purity of a human doing founded merely on itself. It is that which worlds the world, things the thing. Without any opponent to constitute it, any leader to bring it about, this presencing cannot, at bottom, be mastered any more than it can master. It simply is. And so, bringing things forth is not mastering them but revealing their truth. Freedom is not control, submission to a leader, or the conquest of reason's others but is simply allowing truth to happen unopposed. Responding to the governance of the world worlding, we reveal truth when we use our technology to free us from the oppositions that sustained it. We are free when the earth happens unopposed in what we reveal, in what we know.

Between humanity and world a complex interplay occurs. The world's worlding in the gathering of the thing occurs only in the presence of mortal humankind. It is under our care and guardianship, our life and our dwelling, that things are brought forth and interpreted as the things they are. Because of this, the danger that is the truth of modern technology cannot change over into another destiny, another truth, way of being, or interpretation of the thing, without the gentle and undemanding cooperation of humanity.

According to the Thinker, thinking is a handcraft; it must draw near and be near to the thing it thinks, the life it lives, and must let it be.[32] The coming to presence of the destiny and the truth that will spare the earth and free the thing to itself will occur in the lives of the thinkers who will open themselves up to the earth and think another way of being, a way that can be without opposing itself to its other, without demanding the submission of mastery. Their way of being, their technology, and their interpretation of the thing must change and be appropriate for the destiny the earth calls them toward. As this change in their ways occurs, new things—unthought of, undreamed of before because they do not oppose themselves to reason's domain of governance or try to master it—will rise up from the earth and appear in their presence, calling upon them for their thought. The coming to presence of technology as a thing to be thought, the turning to the new world, will occur in a way that restores it into its yet-concealed truth.

As the Thinker suggests, this turning is like what happens when, in emotional terms, one gets over grief or a lost love.[33] The sudden absence of that which was so near to one's life, the loss of links and ties to another, leaves one disoriented for a while, lost to the cares of life. The cares and needs that governed one's life, gathering its actions into meaningful acts, now perhaps poisoned with resentment that absence brings with it, still pull, even though the thing calling them into being no longer is. Everything loses its meaning in the absence of the oppositions that governed the thinging of all things. The world appears as a dream, a distant twilight, but then gradually things draw near again, though in a different way, making possible a different life. And dwelling in the midst of one's cares occurs again.

Moving from one way of Being to another, from a life constituted by metaphysical oppositions to anarchy, from the mastery of leader and follower to letting the world be, humanity must open itself up to the earth, ceasing to live a life built of fear of the other. In keeping with this governing interpretation of the thing, all that is near to the lives of specific, living, and doing people, all that is true to humankind's dwelling place, must first open itself up to the place of technology, the bringing forth of things from the earth. But before mortals can become attentive to the place of technology, before it is possible to deconstruct modernity's oppositions and have a dwelling relation between the truth of technology and the truth of humanity, humankind must first and above all else find its way back to its dwelling place and begin to interpret things as originating in the abyss of Being, the earth, and not as the Reserved for the far-flung imperatives of modern Man's quest for mastery. The dwelling place of man and woman receives its place from the world's worlding, not from reason's opponents, and it is the most true responsibility of humanity to become the world's guardian, to spare the earth on which it dwells, and to let the world world through it.

Unless humanity opens itself up to the earth in all its mystery, ceasing to stand in opposition to it as its master, and there takes up its dwelling, it will not be capable of anything it is called to be. Caught up in opposition to its enemies, it will remain entrapped in the links of modernity. Not seeking Man's mastery, thinking is not governed by the archytecture of reason's oppositions, its fears, its hatreds; it does not know or seek the purity of universal and eternal truths, but rather it brings forth temporal and local truths that are near to the dwelling place, unopposed to its differences. It does not command things forth according to the fatherly logic of its purest origins but rather opens itself up to what appears from the earth, even if it appears in a chaotic dispersion.

Thinking is anarchycal, without a guiding principle to command it, to make it pure, or to oppose it to its enemies.[34] As such, it is radically unlike anything that modernity knows. Before they can dwell as the earth calls them to, to let Being be, to live without the trapped logic of universal opposition, mortals must learn to think. For thinking, according to the Thinker, is an earthy activity, a handcraft that means lending a hand and a care to the earth, in all its dispersion and mystery, as it brings the thing forth. Thinking means building a place for the world to world unopposed, opening up a way for the thing to rise up from the mystery of the earth and become present at. This occurs in the way we live our daily lives, and it is reflected in the language we use. By listening to our language, interpreting the things present in it, we can hear the gentle whisper of the world worlding, and through its calling, we can come to the most radical changes.

True revolutions, real changes in Being, come not with the thunder of ungodly force, the roar of cannons, and the screaming death of those condemned to a judgment of pure evil but on the wings of butterflies and the soft paws of cats, in the gentle murmur of the words we use and the truths we know. And suddenly, without the brutal hand of a leader who would make us different from what we are, the world is different.[35]

Perhaps one day, without the benefit of a leader who would insist on it, the danger that conceals itself as innocent reason seeking only to purify the world of its most evil enemies will, itself, come to presence as evil, a monstrous command of opposition to the evil governing our life, and we will at last begin to think, to build, and to dwell in a way that does not submit to evil, mastery, or opposition. We would then be able to open ourselves up to the anarchy of the earth. No longer imprisoned by the monstrous necessity of modernity, the need to subject everything to the command, to the leader, and to the authority of reason, the thing will be spared, the world will world as world, the earth will be left its mystery, and we will dwell in peace.

Perhaps we who follow the Thinker on his path to anarchy and try to think the thoughts he thought and the ones he left unthought, perhaps we stand already in the shadow cast ahead by the advent of the turning. But we who live with our danger hanging heavy on our shoulders dare not, he tells us, think we can plan out how humanity will dwell without opposing itself to the earth, what technology it will use, what gods or goddesses it will know, nor dare we lay down heavy prescriptions that it should fill.[36] That would be totalitarian, oppositional, hierarchical, and that would make it necessary to purify the world of those who do not fit in this utopia. To imagine how mortals will build after the turn, to dream of an anarchy where the evils of our world are not possible, is to ensure, yet again and more terribly, that our endangered system of metaphysics and opposition prevails. To plan things out as they might be, to dream of utopias and then to try to impose them on the world, is to make them over, ever more securely, into things as they are. That is what our factory planners, our scientific experimenters, our technocrats of discipline have always done. To attempt to save the world from its danger, as the Thinker did when he joined the Nazis, is only to ensure it further, for any such attempt would at bottom rely on that which is bringing it, the truth and reasoning that Technoarchy reveals. Chasing after the future, planning, calculating, and extending the incomplete truths of our time in the hope of creating an order that is not ensnared by our reason and its fears only continues and extends our prevailing attitude of mastering the world through technology and calculating representation.

But this does not mean that we must helplessly submit to our danger, the domain of reason and its enemies, and withdraw from its politics, only that we must be careful about the way we free ourselves from it, oppose ourselves to it. We free ourselves from Technoarchy by thinking of it as a danger sent to us as a whole, by becoming aware of how all the horrors and monstrosities of our time, such as the Nazi death camps, the arms race, the desecration of our land with toxic waste dumps, the collapse of the household, and the slaughter of the whales are possibilities sent to us by the truth we affirm in our lives, our reason and everything we pursue with it. Because of reason's fears, its quest for power and control, its need for making the whole world pure with its absolute presence, all these things have become possible. Reason has built the modern world. Its evils, its monstrous horrors, its opponents and dangers, are not unrelated to its affirmations. Once our truth is present to us as our danger, a monstrousness governing our whole lives, and once we know that it conceals the earth from which it springs, the way is opened to anarchy, a way of bringing things forth without subjecting them to an archytecture that commands them forth in a totality as Man's utility.[37]

Toward the end of his book, Michael Gillespie argues that there is cause for concern in the Thinker's turn toward Being because, according to him:

> We must first prepare ourselves for the experience of Being by purging ourselves of all past metaphysical standards and valuations, of all categories of logic, of all distinctions of natural kinds, of all our conceptions of justice and right, of freedom and necessity, of causality, indeed of every idea, structure, and institution with which we are familiar.[38]

Gillespie's fear as we follow the Thinker in acknowledging the nihilism that is the truth of our world, as we follow Being into the abyss and wait for whatever revelation emerges there and resolutely follow it wherever it may lead, is that we may become monsters, raging agents of destruction and evil. He says: "Having abandoned the categorical reason of metaphysics for something approaching pure intuitionism and the orderly world of everyday experience for the terrors of the abyss, man is thus liable to fall prey to the most subterranean forces in his soul or at least is in danger of mistaking the subrational for the superrational."[39]

This, Gillespie thinks, is why the Thinker was seduced by National Socialism for a brief while. Since the Thinker surrenders all responsibility for his thought to Being, and since there is no ground for distinguishing between good and evil—linked, as they are, as doubles of the same truth—any thinker, Gillespie fears, can practice evil with impunity. As the Thinker himself said, "He who thinks greatly, errs greatly."

Because it goes fearlessly into the abyss, Gillespie argues that the Thinker's thought, despite itself, might be subjective (Gillespie's word), which is to say, whimsical caprice. In thinking, no clear rules allow a thinker to differentiate an authentic revelation of Being from mere caprice, the claims of true prophets from the demagogic claims of false prophets. Thinking does not appeal to intersubjectivity, shared understandings, community standards. Disregarding such standards, it can be caught by anything—the nightmares of madness, the distorted resentment of the weak, the revenge of the wronged. How are the thinkers to know when they are being lead astray?

Quite simply, they cannot, not for sure, but this does not put them in any worse situation than rationalists, theists, or any other archytecture, which usually conceals something at least as errant in its claims to universality, intersubjectivity, or whatever. And thinking has one advantage over any metaphysic of morality: Drawing near to its place, losing all other worlds in its shadow, it knows that it is in error, always and unavoidably, and so is likely to encourage humility, not conquest and oppression.

Besides that, thinking has its way. Silencing her will, opening herself toward friendship, turning toward Being, the thinker attends to the world's worlding and lets truth happen. The way that she reveals things is likely to reveal a world free of fear, suspicion, resentment, domination, or control. This is quite in contrast to the world Technoarchy reveals. As long as she has cause to believe that truth is happening, the thinker has cause to believe that she is not lost in whimsical caprice. She is letting the world world, and in doing so, she sets herself free.

Because thinking does not advance itself with leadership, ideology, or policy, it does not restrict itself to an elite that leads, imposes, or demands. It is something that everyone, whatever their place, can open themselves up to. They think by drawing near to the cares and concerns in their lives, letting truth happen in poetry, song, dance, and prayer while they dwell amid friendship and meditative attentiveness. The truth of the world's worlding cannot be elaborated as a policy that others must submit to, because the criterion of truth is that it happens amid dwelling. It is something people must draw near. There really is no danger, given the way that she approaches truth, that the thinker, so long as she is thinking, is going to fall prey to the most subterranean forces in her soul or mistake the subrational for the superrational, as Gillespie fears—or that thinking will somehow turn her toward Nazism or something like it. Nothing could be more unlikely. The Thinker became a Nazi because he failed to think, because he was afraid, because he lost himself in the they world. It was his failure, not thinking's.

Without an archytecture of any sort, the insecurities of the abyss are indeed frightening, but the horrors of the world that reason has built are worse. Monsters like Sade and the Nazis may well come up out of the abyss in this age, but they can be monsters only because the shadow reason has built is monstrous, made of fear, hatred, resentment, and exclusion. Nihilistic subjectivity, like that of Sade's, is deeply indebted to the archytecture of modern self. Cruel and sadistic whims do not reach far and cannot govern unless a place is already prepared for them. Otherwise they disappear from history, forgotten, unsustainable.

It is reason, with its standards of objectivity, universality, discipline, and the others that it produces to constitute itself, that breeds monsters and prepares a place for them. Having reduced all of humanity to its utility, reason governs the world without allowing any aim or purpose that does not refer back to its arbitrary utility, its groundless and aimless subjectivity. It is reason itself that has produced the monster, the reign of arbitrary subjectivity, that Gillespie fears. Seeking to make the subject's will master of everything, it is reason which destroys limits, ethics, morals, and traditions by subjecting them to its utility and power, by making them into a means. It is reason, not anarchy, which prepares the world for the reign of subjectivity, giving the monster the thought that frees him from any restraint that would stop him from playing with humanity or keep him from making them into objects of his raging utility. And so we need not look at the ways of thinking to find a dangerous thought. Despite its certainty that it is pure, reason is already pregnant with its own monsters.

This does not mean that thinking is not without its dangers, especially not if it is done in the world reason has built. Thinking is the highest form of action, and so, the most dangerous.[40] It is thinking, drawing near to the life which does it, which reveals the monstrousness of our ways and calls on us to adopt others. Building and dwelling according the new truths that it knows makes inevitable the destruction of the old—and all the traditions, institutions, systems, and procedures that depended on them. Even if thinking proceeds in the most gentle and innocent ways—building houses that do not need external heat, finding alternative energy sources, growing food in gardens and greenhouses close to the household—it endangers the economies reason has built with sudden failure because everything was built according to the ready availability of consumer demand.

The Thinker quotes one of Hölderlin's poem in supporting this point: "But where the danger is, grows the saving power also."[41] Thinking this poem more essentially than Hölderlin thought it, the Thinker interprets it as saying that the saving power does not appear incidentally, as random unconnected externality, but as part of the very danger itself, the act of opposition. Modern reason, the evil of our age, is also its saving power, because, as the concealment of the truth of evil by means of its exclusion, it calls on us from concealment to turn to anarchy, to open up to the earth, and to de-construct the archytecture that sustains it. Precisely because Technoarchy conceals itself as eternal reason, it makes it possible for us to open ourselves up to the concealed, to live without the oppositions that bring reason's evil monsters into being. In Technoarchy, concealing is present in opposition, critique, and exclusion.

To save means to free, to spare and husband, to protect and guard, to let the thing thing. The danger that the saving power would save us from is the danger that condemns us to live in a place that is always something other than itself, something at war with itself, a placeless place in which all presencing is determined by a metaphysic that, removed from the earth, is simultaneously everywhere and nowhere. When the truth governing our lives is present as danger, when all about us every-thing is present as something monstrous and evil, a desecration of the earth and known as the Reserved, then, in this moment of deepest despair, the anarchycal nothingness that is the origin of all things rises up and calls on us to think it. As we mortals draw near to this abyss of mystery, silence, and nothingness, we are called on by the world's worlding to speak to the silence and to the no-thingness, to address our thought to what Being, pursuing the modern quest for rational purity, has concealed from us, the monstrous oppositions of Man as will willing only itself.[42]

Thus saved from its own silence, recognized as oblivion, the silence is no longer a silence, a nothingness pure, absolute, and unknown, but the place from which the saving power grows, the place from which things arise unopposed and are no-thing. With such turning, the oblivion that is the destiny of modernity is no longer a nihilism that silences the truth of its own place but rather an anarchycal turning that spares and preserves the abyss of Being as the mystery of the earth. The turning that separates one time from another always occurs suddenly, without explanation, anticipation, or cause. Things come forth from the earth, are present, in an entirely different way, and an entirely different truth presides over whatever is brought forth.

And so, the politics that the Thinker's thought makes possible is not a politics founded on a vision of the future that would justify the exclusion, repression, or destruction of any sort of difference that was not a pure expression of its truth. On the contrary, it would save us from that kind of politics, the kind that built the death camps, that seeks extermination of its others, destruction of opponents. The Thinker had learned something by the time he wrote "The Turning." He had learned that the turning could not be forced with an act of will, brought about by a leader or by raw power making us submit to the spirit, but that, if it came, the turning would be a gentle letting be, a caring and nonjudgmental releasement.

To save the earth, we let the world world and truth happen. We do not build utopias, nor do we invoke universal standards or timeless principles to judge, condemn, or marginalize. We accept the world as it happens at our place and live there in as friendly a way as possible. This way of living is a possibility for a number of movements that are happening in our time. The New Age movement has possibilities in this direction, as does the feminist movement, the peace movement, the environmental movement, and the holistic health movement. Accepting otherness in a way almost no other movement has ever done, the New Age movement often cultivates nonjudgmental releasement toward life, accepting whatever happens as a lesson to be learned.

Feminism naturally reveals the exclusion, oppression, and repression of patriarchy and celebrates what patriarchy has denigrated—sensuality, connection, and physis. If some feminists, like the separatists, have advocated very familiar strategies of exclusion and hierarchy, other feminists have more often tried to break them down. If some feminists have isolated themselves in cults of victimhood and rage, other feminists have reached out toward men and the earth and argued that the woman most in need of liberation is the woman inside of every man and woman. They know that patriarchy is as bad for men as it is for women, even if it is less apparent because men do seem to derive some benefit from it. Men oppress themselves, each other, and the earth as much as they do women, and they will not be free of it until the woman they have cast into their shadow is accepted as a worthy part of themselves and the world. At its best, feminism is not just about gaining equality with men; it is about cultivating a healthy self for both men and women, and a healthy earth to live on.

The peace movement is releasing us from our fear, freeing us of the projections that we have cast on our enemies. It seeks out our shadow, what we are not, and finds ways of becoming friends with it. Not just, at its best, to overcome differences, to conceal and deny them, but to accept and celebrate them.

The environmental movement decenters our human community, revealing the rest of the world, all the plants and animals, as beings belonging in community with us. It upsets the hierarchy between Man and nature and turns us away from dominating the earth to friendship with it. In much the same way, holistic health moves us out of ourselves to our relation with the rest of the world, to the world's worlding. We cannot be healthy, it knows, unless we protect and cultivate the health of everything else. Both the environmental movement and the holistic health movement turn us toward a friendly sparing of all the earth.

The turning is in fact happening; it is happening in each of these movements. They all propose alternatives to Technoarchy, and they all are turning away from Man's willing and domination. Of course, all of them have their moments when they invoke the old metaphysics, but they all are doing it less as time goes by, and they all are becoming more thoughtful. Hope is justified.

CHAPTER TWELVE

Building Wilderness

Mortals dwell in that they receive the sky as sky. They leave to the sun and moon their journey, to the stars their courses, to the seasons their blessing and their inclemency; they do not turn night into day nor day into a harassed unrest. Mortals dwell in that they await the divinities as divinities. In hope they hold up to the divinities what is unhoped for. They wait for intimations of their coming and do not mistake the signs of their absence. They do not make their gods for themselves and do not worship idols. In the very depth of misfortune they wait for the weal that has been withdrawn.

—Heidegger, "Building Dwelling Thinking"

THIS QUOTATION from the Thinker is about how dwelling lets the wildness of things be, how it leaves to the sun and the moon their journey, the stars their courses, the seasons their differences, and the gods their absence. Leaving things alone, dwelling does not impose any truth on the thing that is not its own but lets the wild-erness of Being be. And it does this while it builds a world, while mortals, man and woman, draw things near to their life, handling them, dwelling amid them. Situated in time, life, and culture, dwelling *builds* wild-erness, an anarchycal, centerless, and nonmetaphysical interpretation of the thing's thinging. Forgetting the authority of origins, the claims of patriarchy, and the morality of metaphysics, dwelling cultivates difference, includes alterity, nurtures diversity, protects ambiguity, spares multiplicity, frees irony, and makes it possible to understand it all as the world's worlding.

Before I wrote my dissertation, I returned home from my graduate studies at the University of Massachusetts to my family's ranch in Montana and built an underground house on the south side of a hilltop, as I mentioned in chapter 1. I guess I was homesick. I wanted something to stay near, to be at home with.

The examples that guided my building were the Arks that the New Alchemists had built—greenhouses that nurtured and supported a wide diversity of life, whose boundaries between "inside" and "outside," "wild" and "cultivated," were thoroughly transgressed. Inside, the Arks duplicated, by means of cultivation, the wilderness the outside represented, but lacked.[1] Seeking to live a life that was my own, free of the demands of reason and economy that had ruined so much of the land I was raised on, and wanting a home that I could keep safe, I turned to the Thinker and Henry David Thoreau to guide my thinking. Looking for a home when the whole world seemed lost and homeless, I built to keep my life, my home, and my thought from being claimed by the logic and imperatives of Technoarchy. By situating what it had displaced, by gathering my building around my dwelling, my thought around my life and my place, I hoped to recover from my homesickness. Perhaps Thoreau thought it all for me when he went home to the wilderness around Walden Pond to sort out the irrupting, anarchycal, and possibly mean experience of life from the distant demands modernity had imprisoned it in.

> I went to the woods because I wished to live deliberately, to front only the essential facts of life, and see if I could not learn what it had to teach, and not, when I came to die, discover that I had

> not lived. I did not wish to live what was not life, living so dear, nor did I wish to practice resignation, unless it was quite necessary. I wanted to live deep and suck out all the marrow of life, to live so sturdily and Spartan-like as to put to rout all that was not life, to cut a broad swath and shave close to drive life into a corner, and reduce it to its lowest terms, and if it proved to mean, why then to get the whole and genuine meanness of it, and publish its meanness to the world; or if it were sublime, to know it by experience, and be able to give a true account of it in my next excursion.[2]

Building somewhat more luxuriously than Thoreau because I wanted to build permanently, the earth-sheltered house that I constructed still only cost about $6,000 for materials. With a little help from my family, I did most of the work myself, from designing it, digging out the hillside, mixing and pouring the concrete, erecting the walls, to doing the wiring and the plumbing and the finishing. When I was through a year later, I had a family-sized house and greenhouse that, using only passive solar heat, seldom went below sixty degrees on even the coldest and windiest of Montana's winter days.[3] Unlike Thoreau, I could not feed myself with the food that I grew from my garden because a series of exceptionally severe droughts produced a plague of grasshoppers that have stripped my garden to bare ground several years in a row. Some day I hope to build wind-powered electrical and water systems to make my house completely independent of the utilities.

In this age when the world's industrial economies threaten the entire biosphere with the twin perils of the greenhouse effect and the depletion of the ozone layer, when the ecologically essential forests of the Amazon are being mowed down to provide profits for the fast-food industry, when the topsoil of almost every country in the world is eroding many times faster than it is being rebuilt, when the number of species becoming extinct is comparable only to the Great Extinction that ended the age of the dinosaurs, it seems like a pitifully small thing to build a house that does not need utility heat. And perhaps it is. Perhaps I should be instead engaged in a desperate politics of reform and protest, maybe even revolution. Time is so short. And there is no promise that the Turning will come in time to spare the earth or humanity.

Even though there is desperate need for worldwide change, limited by the knowledge my place makes possible, I hesitate to legislate the law of other places, to provide, at last, a solution to our problems. That, after all, is *the* problem—trying to control things, to separate them out into universal and timeless dualities—this one good, that one bad—and subjecting one to the other. That is the prison our metaphysical archytecture of science, reason, principles, and morals has trapped us in. It has made us homeless by making our body, our place, our culture, our hopes and dreams, and our frustrations and despairs irrelevant and meaningless. They are merely "subjective." Instead of drawing near to our place, we offer universal and objective critiques, judgments, condemnations, and promises of salvation,[4] even when the philosophies underlying them are purely materialistic, as in Marxism or Freudianism.

Seeking universal foundations for action by separating the knower from the known, then building a discursive utopia that supports itself with rejection, marginalization, exclusion, and control, modern critique conventionally leads us to separate our life from ourselves and our place on the earth and put everything under the governance of a politics of cold distance —of rules, systems, laws, principles. Things have to be the same for everyone, everywhere. That is what meta-physics is after all, the submission of the *physis*—Greek for nature or earth—to something above it, beyond it, to something it is not, to reason, logic, principle. This is the patriarchal claim. Building the archytecture of universal reason, eternal truth, and unqualified technology, we reject the anarchy of our place, our home, our life, the earth itself, and put ourselves beyond them. And our lives become governed by things far removed from our own home and our cares, fears, loves, and needs. Caught in the discursive archytecture of Technoarchy, we become homeless.

When time is short and the earth is dying, it seems like we should do something—a dictatorship of ecologists maybe? But perhaps it is especially when time grows short that there is need of careful thought that is freed from the archytecture that has built the systems that, having removed themselves from it, endanger the whole earth. Acting, I suspect, without turning away from Technoarchy's reason and technology and ceasing to live and think will only assure that everything will be rebuilt once again as it was—even under a dictatorship of ecologists. If it survives the archytecture of Technoarchy, humanity will build and live differently, will think first of home and act first locally. It will learn to respond with care to the earth, the sky, the sacred, and the death of all mortals, instead of a reason that is far removed from the earth. The thoughts of Technoarchy must be deconstructed where they are the most secure—in the lives of we mortal humans who build and dwell, in our dependence on its institutions and its systems that trap us in its networks of technopower, its bipolar hierarchies, and the reign of monstrous others it produces.

Freedom does not come by making the world conform to our prescriptions, demands, and chosen imperatives, but only with a gentle releasement toward Being, with a meditative listening to the whisper of the world worlding as it happens at home. Then, as an-archysts repudiating the archytecture of reason, ideology, technology, methodology, and principle, we build and dwell in a way that does not destroy the earth. Anarchy is the way of Being following the Turn, the way in which the thinker lets the wilderness of things be.

Perhaps this is in its own way utopian, that is, inappropriate to the dangers at hand. (If it is, it would not be the first time that I have preached against the sins I am most guilty of.) Perhaps things have become so horrible that this can be described as a retreat, a withdrawal from the world, and that, by letting this Being be, it simply consents to letting the nightmare continue. Perhaps so. I cannot be sure. But I believe that it helps. Whether they are able to make a change in the world or not, anyone's effort at thinking, at drawing near to their place, caring for it, makes the world better because it lets truth, friendship, and freedom happen. I cannot offer a theory of why such local efforts can have any effect of importance in the world; I can only hope and trust, even when these efforts seem to touch nothing but themselves. But I digress.

The wilderness, or anarchy, of Being is not the opposite of civilization, as it has long been characterized in the Western tradition—virginal, unhandled, inhuman, untouched—but rather a building that we dwell in, that we have built because of what we are. In Being's wilderness we do not strip away our connections, our belonging with others, becoming a lonely outcast in the world's vastness, at last free of ourselves; rather, we find a place where we learn of our life's connections with otherness, with the shadow we think we are not, with the community of Being we have tried to escape. In going into the wilderness, which is as easily found in the city as the vast rain forest, we are going home because wilderness is the place where we recover the things that are most ourselves but that we have denied, repressed, and forgotten. Building wilderness is a lot like interpreting dreams. In doing it, we encounter the surprise of otherness, a shadow that is not really so other because it is our own being. A returning of the other, it is a place in our life that reminds us of our ties to the earth and our place on it. As such, it is a life governed by the inner and situated ways of thinking and love, not by the external and distant summons of patriarchy's morals, principles, or reason. To dwell is to build a place to think of love, care, and peace.

Surprisingly, the Old English and the High German word for building, *baun*, means "dwell," "remain," "stay in a place," according to the Thinker.[5] The original meaning of the verb *bauen*, "to dwell," has been more or less lost to us. But in the word "neighbor," in Old English *neahgebur*, a trace of it remains. *Neah* means "near," and *gebur* means "dweller"—near-dweller. Not only does the old word *baun* tell us that to build is really to dwell, it also suggests what dwelling brings forth—Being. In German, the old word *bauen* is also related to the words *bin* and *bist*. Thus, *ich bin*, "I am," and *du bist*, "you are," mean, not only that we are, but that I dwell, you dwell. We live; we have a place that we live at, and the world worlds because we are there. The way in which you are and I am, the way in which we, as mortal beings, are upon the earth, is as dwellers. To be a human being is to be a dweller, a human whose life is built amid a place on earth. Living, we build a world for ourselves, tend for it, care for the people and things that share it with us. Even such things as "nature," "the gods," "humanity," and "death" are buildings, names for the thoughts that we dwell amid, construct our world with. The world worlds when we build, dwell, and think.

Building also means, though less commonly now, to cherish and protect, to bring forth as a preserving and caring, and especially as a cultivating and nurturing of the earth. For example, after many years of careful work, after a farmer has cultivated a rich layer of humus in her soil, she says that her soil is built. As cultivating and nurturing of the abyss, building is an open and responsive caring that brings forth the gods. Making possible any interpretation of the world's worlding, it takes place before them and seeks their blessing and their gifts in a bountiful harvest. To practice the art of agriculture, as the past of the word suggests, is to cult-ivate the favor of the gods, to bring their message forth, to attend to the earth that conceals it, and to abide with the truth brought forth.[6]

But not all building is tending to the soil, since ships and temples are also built. This distinction does not mean, however, that building as cultivation and building as construction are two different things, contraries that must oppose each other. As ways of making things, both modes of building bring forth, cultivate and care for, the sacred ones that govern dwelling.[7] All thoughts, names, words, or things are buildings, care-ful constructs made for the purpose of dwelling. Building is always world building, and whatever is built is done in poetry, prayer, and song.

Even in this, the darkest of times, when the gods have fled and poetry is an embarrassment, prayer a superstition, and song an industry. When Technoarchy darkens the world, building as artful cultivation is eclipsed, and building as willful fabrication guided by the universal imperatives of technological efficiency comes to the foreground, concealing the world's mystery and beauty. "The earth and its atmosphere become raw material. Man becomes human material, which is disposed of with a view to proposed goals."[8] This is a decisive occurrence: dwelling is no longer experienced as humanity's being, its way of living in the world.[9] Instead, harnessing the whole world to its cold and de-secrating logic, it becomes a slave to Man's distant economies and imperatives, and the wild anarchy of the world's worlding is concealed. Even so, this silence that conceals the poetic and sacred character of dwelling can yet be listened to, heard beneath the distracting noise of modernity's archytecture of definitions that it builds in the service of Man's reason.

If we cultivate this silence, this flight of the gods, the Thinker says, we can still hear the calling amid our life's cares that calls us to think building as dwelling, to spare, venerate, and free the wildness in our being on earth, and to understand building as a cultivating of the abyss. Without foundation, archytecture, or universal reality, the abyss is the true ground, the earth on which our world is built. It is necessary to acknowledge this in our thinking before our building is free:

> The word for abyss—*Abgrund*—originally means the soil and ground toward which, because it is undermost, a thing tends toward. But in what follows we shall think of the *Ab-* as the complete absence of ground. The ground is the soil in which to strike root and to stand. The age for which the ground fails to come, hangs in the abyss. . . . In the age of the world's night, the abyss of the world must be experienced and endured. But for this it is necessary that there be those who reach into the abyss.[10]

If we reach into the abyss that now conceals the poetry in building, we come to understand that we do not dwell because we have built, have erected houses, bridges, and roads, but we build and have built because we dwell, because we cultivate the (absent) gods in our thinking, because we are, as living mortals, possessed by the summons of poetry and prayer as they rise up singing from the abyss.[11] Just as we cannot speak language, master all its ambiguities and subject it to our will but must yield to its appropriation of our being, allowing it to speak us as the callings we are, so too we build only because we dwell, only because the song of poetry has already gathered us into the absence that governs the thinging of the thing.

"But in what," the Thinker continues, "does the nature of dwelling consist?"[12] To answer we must again follow language back into its home. According to the Thinker, the German word for "dwelling," *wohnen*, has its roots in the Old Saxon *wuon* and the Gothic *wunian*, which both, like the old word *bauen*, mean "remain," "stay in place." But unlike the word for dwelling that later developed into the English word for dwelling, the Gothic word is more descriptive of how this dwelling is experienced. *Wunian* also used to mean "be at peace," "be brought to peace," "remain in peace." As Old English did once too, the German language said "peace" with the word *friede*, now meaning, "the free." Long ago, the word "free" was associated with what was loved and called for protection from harm and danger, safeguarded in its nature. To free means to spare, love, befriend, and care for—to spare not only in the negative sense of not harming what we love, of not setting upon it as a means, a tool, a way to will the will, but also, and more important, in the positive sense of leaving something beforehand to its own nature, actively, thoughtfully, lovingly preserving it in its peace, keeping it safe in its serenity or tranquillity. Leaving their being wild, in other words, free of any archytecture that would deny them their truth.

To dwell, then, in its most profound sense, is to preserve things in their peace, to spare them actively from anything that might disturb them or make them different from what they are, as a lover would a beloved, a mother her child. As dwellers, our calling, as the Thinker has described it from time to time, is to be the "Guardians of Being," the friends of the world's worlding, the lovers of the earth's wilderness. The fundamental way of dwelling, even in this destitute time of the world's night, is this nurturing sparing and preserving that accepts things as they are—despite their wildness, their difference, their contrary nature—and allows them to become what they will.

Gently unassuming, making none of patriarchy's claims, dwelling builds no centers that demand the submission of everything in the world to principle, ethic, or law. And so it is not anthrocentric, or even biocentric or ecocentric. The haunting wail of the coyote, the timid wanderings of the rabbit, the predatory hunger of the wolf and the bear, the graceful leap of the deer, the awesome complexity of the whale's song, the formidable hiss of the mountain lion, the soaring arch of the eagle—all of these are buildings that house the world's differences. And because the world is built of them, the thinker loves all of them as they are, without judgment, evaluation, or reservation. And she will in turn build a way to dwell in peace with them. Throughout dwelling's whole sundering breadth the acceptance of the friend, the foresight of the wise, the care of the lover, and the courage of the hero pervade dwelling, determining its whole way of being, which is the stay of mortals on the earth, under the sky, before the gods. To dwell, the Thinker argues, is to gather things together—the earth, the sky, the gods, and the life together of mortals—and to accept each, however different, however other, as a part of the dwelling place, keeping safe the peace of the world.

The earth, the sky, the gods, and the mortals are the different aspects of the world's worlding. Worlded by the world's worlding, mortals dwell by sparing and keeping safe the oneness and the alterity of earth and sky, gods and mortals, that happens in the fourfold occurrence of things. As the Thinker says: "Thinging, the thing stays the united four, earth and sky, divinities and mortals, in the simple onefold of their self-united fourfold."[13]

As the fruitful womb of all that rises forth as plant and animal, that irrupts with rock and water and yet takes it all back again in death, decay, and time, the earth is the wild and concealing darkness, the fertile mystery hiding the truth of things in dark obscurity.[14] Says the Thinker, "Earth is the building bearer, nourishing with its fruits, tending water and rock, plant and animal."[15] Earth is what the early Greeks thought as chaos. Chaos, according to Vycinas, was not the mindless disorder that we think now under the reign of reason's archytecture but the open abyss, the wild nothingness, the groundless ground, the womb from which things rise up and appear of their own mysterious, unknowable nature.[16] Physis, as the mystery and power of the earth that brings things forth, brings them forth from the earth's concealment.[17] For the Thinker, the earth, as physis, is not an object of utility, a universal archytecture, or a timeless foundation but a *way* of Being, the way it is when it is concealed. Despite its unfathomable anarchy, it, like the other parts of the fourfold, is a building, a construct of dwelling and thinking brought forth by the world's worlding. The source of the wilderness of being, the earth keeps and safeguards the seeds of things that, in their own way and time, rise up into the sky to be greeted by the gaze of mortals and drawn near to the bounds of the holy.[18]

The sky, as the horizon surrounding the place of mortals in their life upon the earth, reveals things as they present themselves in the life of mortals. "The sky is the sun's path, the course of the moon, the glitter of the stars, the year's seasons, the light and dusk of day, the gloom and glow of night, the clemency and inclemency of the weather, the drifting clouds and the blue depth of the ether."[19] All these things that appear under the sky appear in the nearness of mortals. The world worlds the nearness of their life, the travails, movements, and gods they pray to. Gathered beforehand into the world's worlding, the horizon changes with their wanderings, and things rise up in their presence and fade into obscurity with their passing.[20]

According to the Thinker, "Mortals dwell in that they receive the sky as sky. They leave to the sun and moon their journey, to the stars their courses, to the seasons their blessing and their inclemency; they do not turn night into day nor day into a harassed unrest."[21] The way the world worlds when it is unconcealed, the sky is sky only because it stands in contrast to the earth, its other. What the earth brings forth, the sky reveals. The alterity between earth and sky is the contrast between Being as the abyss and Being as presence, between concealment and unconcealment. Dwelling is letting be this otherness, this difference.

The gods are the buildings, the truths, that gather mortals into their care, the messenger-bearing myths that surround and govern the life of mortals with the reality of their dwelling. They are the bearers of the holy, the truth, which bounds the comportment of mortals to the things that appear under the sky, on the earth. Says the Thinker, "The divinities are the beckoning messengers of the godhead. Out of the hidden sway of the divinities the god emerges as what he is, which removes him from any comparison with beings that are present."[22] Even in a destitute world, a time marked by the flight of the gods, the absent gods still, by being the other to mortality, provide the limits and truth of mortal relations to things—how they bring them forth, how they use them, even what they are and how they are known. The flight of the gods means that mortals cannot have a free or peaceful relation to things, but yet things still are, which means that the absent gods still govern the way that we cultivate things, build worlds, if only in their absence.[23]

According to Vycinas, a god for the Thinker is not as we Christians know it, a supernatural entity that reveals its presence in a miraculous or unnatural intervention. Instead, the gods are more like the early Greeks knew them, as truths that are revealed in the ordinary way of things, never against them. For the Greeks, a god does not have to disturb or distort nature to be known, but rather is known as the truth of nature. A god is not above nature, outside it, or apart from it but is in it and finds its way through it. As a result, Greek thinking (before Plato) did not de-secrate nature as Christianity has since Augustine, withdrawing the holy from it, but rather it revealed the holy through it.

A god is a world, a building that gathers things into it, providing them their place and interpretation, even when they present themselves in sundering or dispersion. Because the Greeks knew many gods, they knew many worlds or ways of being and were present in them all simultaneously—ambiguously, tragically, ironically. Because the truth, the way of Being, or interpretation surrounding it is different, the thing is different in each world because it houses a different god.

Night, for example, in the world of Artemis and night in the world of Hermes are different buildings because they disclose different truths. Since Artemis is the goddess of wild and unexplored nature, her nights are frightful and mysterious,[24] and since Hermes is the god of luck, thieves, and gamblers, his night is an advantageous or disadvantageous cover for one's pursuits.[25] Similarly, love, as a truth of Hermes, is a matter of luck or opportunity, a pleasant occurrence because it brings nothing with it but itself, while as a truth of Aphrodite it is a blissful unification, breaking all bounds and inviting sudden tragedy because it overwhelms conventions, ethics, and responsibilities, bringing with it scandal and disgrace.[26] Gods do not specialize in a portion of nature—Apollo the sun, Aphrodite sex—as we Christians think; they house all things in the world, and they reveal their mystery through them all.

A god for the Thinker is the gathering, the building or the circling boundary, which holds things in their place, making possible an interpretation of them. A thing is because it houses a god; it is what it is because it houses a particular god, even though its nature may be something entirely different if a different god is invoked. "Nature" was never a fixed archytecture or foundation for knowing things but a building that was continually remade as the god housing it changed. As a result, for the Greeks the reality of the gods never hinged on their power as causes because they were known not as nature's puppet masters but as strife-torn truths bringing forth the realities of things. The gods were nature—in all its ambivalent modes.

While the gods are the truths of all things, mortals are the measure of all things, their death the possibility governing the building of all things—even the gods and nature. Their mortality, their death, which, as a terminal nothingness, beckons to them all their life long, brings them into the world, forcing on them a need for building that moves them near to the things and makes them responsive to their irrupting wilderness. Says the Thinker, "The mortals are the human beings. They are called mortals because they can die. To die means to be capable of death *as* death. Only man dies, and indeed continually, as long as he remains on earth, under the sky, before the divinities."[27] Summoning them to their lives in anxiety and dread, death gathers mortals up into the world, situating them, limiting them, compelling their attention, making their life whole, holy, complete, meaningful by making them accept their calling as the Guardians of Being.

For mortals, death is the shrine of nothing, an empty void that nevertheless is, that, in its nullity, in the care it imposes and demands in our life as dwellers, governs the presencing of things. As something radically other to it, it presents the mystery of Being itself, the astonishing presence of the thing, and it is the measure of every mortal life, the conclusion that calls us to our truth, to accept the summons of dwelling and thinking. Death governs our building, situating us in our limits, and setting us to our work. But even while we are building, says the Thinker, "man is allowed to look up, out of it, through it, toward the divinities. The upward glance passes aloft toward the sky, and yet it remains below on the earth. The upward glance spans the between of sky and earth. This between is measured out for the dwelling of man."[28] The world and everything in it is as it is because as mortals build things they anticipate the nothingness of death and against it ground the presence of the thing as a reality in their life. The earth's erupting finality, death governs all humans, confining them to the sky's horizon, under which things rise up from the earth, bounded by the sway of the gods. Death calls on humanity to dwell upon the earth, under the sky, before the gods.

Anticipating their death, caring for their life, and building the world, mortals dwell in saving the earth, setting it free. They do not seek to master it, to subjugate it, to tame its wild ways, in order to make it a slave to their thoughtless whims. Mortals dwell in receiving the sky as sky, and they dwell when they let the earth bring things forth in its own way. Mortals dwell in acknowledging their situation and their limits and in accepting death as their fate and living life in the face of death.[29] Mortals dwell in seeking a good death, a life that matters. But this does not mean that they darken their days with gloomy meditations on their end, nor does it mean that they make death their goal, argues the Thinker. The prospect at which the earth possesses all humans, body and soul, death is the guide of mortal life, the place where it erupts and asserts its sway, suggesting how things will present themselves. It is the guiding concern that determines how a thing is to be done, whether it is to be done or left undone. Death is the shrine of being, and mortals dwell through acknowledging its presence in their lives, through accepting the cares it imposes.

Dwelling occurs in saving the earth, in receiving the sky, in awaiting the divinities, in facing death. It is the fourfold preservation and sparing of the fourfold, the setting free of the thing into its own differences. Dwelling preserves the thing as it rises from the earth, and mortals dwell in letting things be in their presencing. Mortals do this by nursing and nurturing the things that grow and carefully constructing the things that do not grow. Seeking to secure the fourfold in things, to let nature be itself, mortals build, bringing forth in their cultivating and constructing their dwelling place.

In the second part of "Building Dwelling Thinking," the Thinker gives some examples of what the relationship between building and dwelling is. The first is of bridges, which are buildings that gather a whole world together—the economies of life, the beauty of the sky, the changes in the weather, the connections between places. A bridge is a building that unites the fourfold. He says:

> Always and ever differently the bridge escorts the lingering and hastening ways of men to and fro, so that they may get to other banks and in the end, as mortals, to the other side. Now in a high arch, now in a low, the bridge vaults over glen and stream—whether mortals keep in mind this vaulting of the bridge's course or forget that they, always themselves on the way to the last bridge, are actually striving to surmount all that is common and unsound in them in order to bring themselves before the haleness of the divinities.[30]

A crossing linking earth and sky, the divinities and the mortals, the bridge is a thing; it gathers the fourfold, and in doing so, it allows a site for the fourfold, a place for it to come to presence. A river has many points that can be crossed by a bridge, but until a bridge is built, none of them are places, spaces within which something has been made room for. A place is something that has been cleared and set free, something built, bounded by a boundary—revealed. But for the Greeks, and for us seeking to find the way all things are built, a boundary is not the geometric point where something ends but the beginning where something begins its presencing, not an encircling mark imposed by the will but a horizon for revealing the thing.

Space is space because mortals dwell in it, handle it with their hands and their tools, bringing forth the god that houses it. Thus, spaces receive their being from their locations, which are locations because of their place in the lives of mortals, not from geometric space itself, an abstraction beyond the life of humanity.[31] Even spaces preserved from the direct hand of Man are opened up as spaces by their proximity to human life—the Custer Battlefield, the Medicine Rocks State Park, the Custer National Forest. Spared from industrial development, these spaces of wilderness receive their location, their boundary, from the existence of Man. Wilderness, for Technoarchy, is merely something Man is not, something he has not imposed his will on. Yet even here, wilderness is something built because Man designates it as his other, thereby bringing it into being. Despite Man's depravity, his boundless will to define everything as his utility (if only by exclusion), something sacred, something beautiful, something magical still speaks from these locations and protects these things from the hand of the developer, the strip miner, and the businessman.

The nature of building is letting dwell, of accepting life on the earth under the sky and responding thoughtfully and carefully to their summons. The Thinker gives an example of this in an old farmhouse in the Black Forest, built two hundred years ago.[32] In this house the self-sufficient power of dwelling let earth and sky, divinities and mortals, enter into simple oneness and ordered the house into being.

> It placed the farm on the wind-sheltered mountain slope looking south, among the meadows close to the spring. It gave it the wide overhanging shingle roof whose proper slope bears up under the

> burden of the snow, and which, reaching deep down, shields the chambers against the storms of long winter nights. It did not forget the altar corner behind the community table; it made room in the chamber for the hallowed places of childbed and the "tree of the dead"—for that is what they call a coffin there . . . and in this way it designed for the different generations under one roof the character of their journey through time.[33]

The farmhouse was built only because the craftsmen who built it already dwelled, respecting the weather in its changes, knowing both birth and death, and responding to the needs of life.

The housing crisis, the homelessness that we face today, comes not from any lack of housing, though there is that in tragic plenty, but from modernity's rootlessness, aimlessness, and nihilism—from our life being rooted not in the earth but in some cold distance. The homeless that fill our streets—dirty, unkempt, cruelly tousled by a society of affluence—are only a symptom of a deeper homelessness that we all share with them and makes us indifferent to their life of despair and pain, a deeper failure to spare the earth. Subject to the utility of a global economy, people cannot dwell in trailer-houses, awaiting their next job a hundred miles away. Cut off from the earth that sustains them and the family that gave them birth by a god that is no god, they cannot have a care for the earth they live upon, nor can they offer a thoughtful interpretation of what anything is for; they can only drain it for what they can use and move on when the smell becomes too strong and the poison too deadly. Denied initiation to their nature as mortals, subjected to their utility as instruments of the modern economy, they cannot accept their ties to the earth and thus come to care for it, set it free.

Before anything can be spared, mortals will turn from its way and build not by the standards of Man's reason, which closes itself off from the earth and builds according to the homeless archytecture of Man's will; rather, mortals will build by beginning to think, by opening themselves up to the earth and responding to the world's worlding. Handmade, near to the life of mortals, building as thinking responds to the dwelling place—its climate, its soil, its people, and its gods. Accepting its situation, it builds as a response to the world that worlds there and leaves the things present themselves there as things deserving the care of thinking. Celebrating the care of friendship, the nurturing concern of the lover, it does not reduce the wilderness of Being to Man's utility, and it does not subject it to necessities of distant and thoughtless meta-imperatives.

A discursive prison that has locked all the earth up in its archytecture and made it available as Man's utility, Technoarchy builds homogeneity over vast areas, callously leveling everything over, bounding it, and organizing it according to an indifferent plan, a distant and calculating will. To secure Man's willing, it makes everything into a transparent re-presentation of Man's utility, controlled, secured, defined. The disorder, dispersion, dirt, noise, irrationality, ambiguity, and slack that are the shadow of Man's building the world as his utility are rigorously excluded because, being wild and uncontrolled, they are not Man's will and threaten the security of his intentions. Such wilderness is dangerous and must be bounded, isolated, separated, and distinguished from Man's civilization. And if it is there "protected," as in parks, it is only to make it safe by distinguishing it. This is not Man's domain.

Thinking, on the other hand, responding to the builder's time and place and not to a distant summons of an inappropriate technology or a global economy, builds to protect heterogeneity, difference, diversity, and dispersion—to protect the wild-erness of the dwelling place. For thinking does not seek to bound wilderness up as something uncontrolled and distinguish it from Man because it does not know that difference.

In diversity, anarchy, and the openness to it, the mystery of the earth and the holiness of the world's worlding are revealed. Thinking builds to spare the wild-erness of things not because it is useful to do so but because in its sparing it is opening itself up to the happening of truth. Thinking knows that truth is not Man's truth, an ordering that is projected over things by Man's will, but is the world's worlding, the happening of the thing at a particular dwelling place. Thinking is directed toward the thing in all its wildness, and because it is, it does not seek to fabricate Man-made simplicity by positing a boundary of control, imposing operational definitions within it, and using analytic methods to extract Man's truth from it. Thinking can know its truth and build according to it only when things are left alone, left wild, when they are not made into objects of Man's will but are accepted and allowed to go their own way. Then, and only then, the whisper of the world's worlding can be heard. Seeking truth, the thinker must silence her willing and, in that silence, let the earth come forward.

Despite thinking's being a response to the local situation, it is possible to hint at some things thinking would tend to build and the way it would proceed. Since thinking does not seek control over vast extents—and indeed cannot because human control over a thing's presencing conceals its truth—it will not build vast energy grids but will instead build local, small-scale, and simple energy systems because they will be responsive to the nature of local needs and the handmade cares of the household.

Instead of relying on distant economies to supply it with food, housing, energy, and many tools, the thoughtful household would attempt to make most of the things it needs itself. And instead of depending on a global economy to supply it with things it cannot build itself, it would rely on barter and community work.

Perhaps in this time of gathering storm clouds the buildings that thinkers would build most often are arks, secure places that protect the wild-erness of things from the callousness of Man's utility. In biblical times, Noah built an ark to give protection to two of every kind of creature, one male, the other female, sparing them from the rage of God's judgment against humanity's corruption.

Whatever it is in the patriarchal biblical tradition, thinkers responding to the signs of the gathering storm might find that the ark can be used as an instrument for protecting anarchy, for preserving difference, for freeing the thing to thing, for building wilderness. A strong place made to keep things safe, it does not seek to subject the things within its protection to command, control, or an archytecture of any sort, but to spare them from the tempest that rages outside it, keeping them safe for a day when they can leave its confines and freely go their way. Since the time of Noah, the ark has been a holy instrument, a sacred instrument because it protects the whole dispersion of things and keeps their integrity safe. It was in an ark, an ark of an entirely different construction, we should be reminded, that the holy objects of the Jews were kept safe from the trials of time and use.

Noah's ark was built to protect the creatures of the earth from the flood God judged the world with, but ours will be built to protect the sacredness, integrity, and anarchycal dispersion of the creatures of the earth from the careless and indifferent judgments of Technoarchy. Man's utility has covered the earth with a monoculture of plants that he has judged the most productive for his economies, pushing all others to a shrinking margin, specially designated "wilderness" areas, where they are endangered because the ecosystems that nurtured them are falling apart in the space left to them. Following the broad diversity of the earth's plants to the margins of Man's utility, many species of animals, such as the wolf, the bear, the whale, and the mountain lion, are similarly endangered. Amid the vast fields of Man's monoculture, many insects become pests and are treated with heavy doses of poison, further disrupting the cycles that renew and protect the earth.

As a result, the task of modern ark builders, responding to the looming storm clouds of the world they dwell in, will be to find ways to spare the anarchy of these dispersed beings from the tempest of Man's will. The seeds of endangered plants will be gathered and planted, animals endangered by a collapsing ecosystem will be gathered up and placed in zoos that nearly as possible duplicate their usual habitat, protected for a day when they can again resume their way. In addition, libraries, sanctuaries, institutions, and communities will be built to protect and nurture dissident ideas, places where poets and thinkers can renew themselves, attending to the wilderness within themselves that alone makes possible a renewal of the earth. One example of this kind of ark building is William Connolly's pursuit of an "agonistic ethic of care for otherness." Resisting the effort to build the world of evil others, Connolly suggests that we cultivate an ethic of care for the ambiguities involved in any affirmation of identity or difference.[34]

As in Noah's time, ark building does not begin with large numbers of people, political action, tight organization, or extensive plans, but with a thoughtful and personal response to the perils of the time. While nothing can be done by reason unless everyone does it together, thinking begins in a small way with the individual opening herself up to the whisper of the world worlding at her place, and it proceeds with the responses she makes in her building and dwelling. Since it builds small and comes to its truth through the things near to it, thinking can, as it always must, start with one thinker dwelling amid the cares of her own life, quietly meditating on the wilderness of things most near to it. By cultivating friendship, protecting difference, nurturing growth, and accepting challenge in her own life, the thinker builds a wilderness, and once this world is built, it can be shared with others.[35] Perhaps by living this gentle life, without invoking a metaphysics of principles, condemning judgments, or universal morals, she can open up ways for others to live their lives in a healthier way.

Notes

1. Thinking about My Place

1. For an excellent description of a farm community under stress, see Jeff Pearson and Jessica Pearson, *No Time But Place: A Prairie Pastoral* (New York: McGraw-Hill, 1980), p. 258. The way the Indians were pushed from their home, the land, is one of the ugliest pages in American history. See Dee Brown, *Bury My Heart at Wounded Knee* (New York: Pocket Books, 1981).

2. U.S. Department of Agriculture, *The 1984 Fact Book of U.S. Agriculture* (Washington, D.C.: U.S. Government Printing Office, 1984), p. 1.

3. Ibid., p. 2.

4. Ibid., p. 82.

5. Ibid., p. 3.

6. Ibid., p. 4.

7. Ibid., p. 2.

8. Ibid., p. 1.

9. U.S. Bureau of the Census, *Statistical Abstract of the United States, 1982–83* (Washington, D.C.: U.S. Government Printing Office, 1982), p. 467. The size of the food budget varies with income—29.7 percent for low-income families, 24.1 percent for intermediate, and 19.4 percent for high in 1981.

10. Michael Best and William Connolly, *The Politicized Economy* (Lexington, Mass.: D. C. Heath, 1982), p. 54.

11. That the work of mostly women in the household is not included in the gross national product is an outrage that indicates a deep contempt not only of women but also of work that is not done in the service of capitalism. Wendell Berry makes a brilliant argument about what kind of work really matters in "The Body and the

Machine," *Parabola: The Magazine of Myth and Tradition* 15, no. 3 (August 1990): 66.

12. Wendell Berry makes this argument in many different places, for example, in *Recollected Essays, 1965–1980* (San Francisco: North Point Press, 1981), p. 157.

13. In 1930, farmers composed 24.9 percent of the population; in 1981, only 2.6 percent (U.S. Bureau of the Census, *Statistical Abstract of the United States*, p. 649).

14. The distinction between thinking and reason and the extended interpretation of them comes from Martin Heidegger's book *What Is Called Thinking?* trans. J. Glenn Gray (New York: Harper and Row, 1968). Although there are many similarities between Heidegger's thinking and Derrida's deconstruction, one difference seems to me to be important. While deconstruction is an excellent strategy for attacking constructions, creating the possibility for space and slack within discursive practices, it is not a possible strategy for sustaining life, for building a world of nurturing care for the earth. Deconstruction is a weapon, a relentless strategy for subversion; thinking is a way of life. This does not mean that deconstruction is evil, only that it is a tool limited to specific purposes, that of attacking the prisons that metaphysics builds.

15. Martin Heidegger, *Discourse on Thinking*, trans. John M. Anderson and E. Hans Freund (New York: Harper and Row, 1966), p. 46.

16. Ibid., p. 45.

17. Ibid., p. 47.

18. From *The American Heritage Dictionary of the English Language, New College Edition.*

19. Heidegger, *Discourse on Thinking*, p. 47.

20. Ibid., p. 47.

21. On how academics handle their "others," see Ellen Schrecker, *No Ivory Tower: McCarthyism and the Universities* (New York: Oxford University Press, 1986), p. 265. Reason is not ashamed to use the most irrational methods to silence, exclude, or marginalize its critics. Seldom have I ever heard rationalists give *reasons* for their dismissal of postmodernism, and when they do, it almost always betrays deep misunderstandings of it. The reign of modernism, scientism, positivism, rationalism (or whatever) is entirely, by its own standards, illegitimate because it rests not on the foundations of logic, open debate, and appeals to evidence but on naked power, censorship, close-mindedness, and a stubborn refusal to encounter

its critics. Defenders of reason deserve to suffer the torments of a very bad conscience, yet there are few indications that they do. Nor are they likely to anytime soon; they are so craven. I personally think that rationalists have sex with animals and dead people every full moon, but I lack empirical validation—so far.

22. See Hannah Arendt's *Eichmann in Jerusalem: A Report on the Banality of Evil* (New York: Penguin Books, 1971).

23. The need for command centers to be able to control their forces in order for deterrence to work is argued in Bruce Blair, *Strategic Command and Control: Redefining the Nuclear Threat* (Washington, D.C.: Brookings Institution, 1985), p. 65.

24. On how reason first posits its other then deploys a fear of it in order to isolate it, see Michel Foucault, *Madness and Civilization*, trans. Richard Howard (New York: Vintage Books, 1973), p. 199.

25. Ralph Metzner, "Getting to Know One's Inner Enemy," in *Human Survival and Consciousness Evolution*, ed. Stanislov Grof (Albany: State University of New York Press, 1988), p. 50.

26. For a discussion of deterrence theory, see Lawrence Freedman, *The Evolution of Nuclear Strategy* (New York: St. Martin's Press, 1983).

27. See Timothy W. Luke, "What's Wrong with Deterrence? A Semiotic Interpretation of National Security Policy," in *International-Intertextual Relations: Postmodern Readings of World Politics*, ed. James Der Derian and Michael Shapiro (Lexington, Mass.: Lexington Books, 1989), p. 207.

2. Thinking about Technology

1. Martin Heidegger, "The Question concerning Technology," in *The Question concerning Technology*, trans. William Lovitt (New York: Harper and Row, 1977), p. 3.

2. Ibid., p. 4.

3. Ibid., p. 5.

4. Ibid., p. 4.

5. Ibid., p. 5.

6. Ibid., p. 6.

7. Ibid., p. 8.

8. The following argument about the causality of the thing comes from Heidegger's "The Thing," in *Poetry, Language, Thought*, trans. Albert Hofstadter (New York: Harper and Row, 1971), p. 170.

9. Ibid., p. 174.

10. Ibid., p. 175.
11. Ibid., p. 177.
12. Ibid., p. 178.
13. Heidegger, "The Question concerning Technology," p. 12.
14. Reiner Schurmann, *Heidegger on Being and Acting: From Principles to Anarchy*, trans. Christine-Marie Gros (Bloomington: Indiana University Press, 1987), p. 171.
15. Ibid., p. 98.
16. Ibid., p. 102.
17. Riane Eisler, *The Chalice and the Blade: Our History, Our Future* (San Francisco: Harper and Row, 1988), p. 30.
18. Ibid., p. 75.
19. Ibid., pp. 32–33.
20. Ibid., pp. 36–37.
21. Ibid., p. 38.
22. Ibid., p. 61.
23. Ibid., pp. 66–73.
24. Dorothy Dinnerstein, *The Mermaid and the Minotaur* (New York: Harper Colophon Books, 1977), p. 80.
25. Jonathan Culler, *On Deconstruction: Theory and Criticism after Structuralism* (Ithaca, N.Y.: Cornell University Press, 1982), p. 59.
26. Ibid., p. 61.
27. Martin Heidegger, "On the Essence of Truth," in *Martin Heidegger: Basic Writings*, ed. David Krell (New York: Harper and Row, 1977), p. 117.
28. Ibid., p. 120.
29. Ibid.
30. John Rawls, *A Theory of Justice* (Cambridge: Harvard University Press, 1971), p. 395.
31. Rupert Sheldrake, *The Presence of the Past: Morphic Resonance and the Habits of Nature* (New York: Vintage Press, 1989), p. 11.
32. Ibid., p. 12.
33. Heidegger, "On the Essence of Truth," p. 121.
34. Martin Heidegger, "The Origin of the Work of Art," in *Martin Heidegger: Basic Writings*, ed. David Krell (New York: Harper and Row, 1977), p. 175.
35. Heidegger, "On the Essence of Truth," p. 130.
36. Ibid., p. 132.
37. Heidegger, "The Origin of the Work of Art," p. 172.
38. Ibid., p. 181.

39. Heidegger, "On the Essence of Truth," p. 138.

40. Fred Dallmayr, "Ontology of Freedom: Heidegger and Political Philosophy," *Political Theory* 12, no. 2 (May 1984): 228.

41. Hannah Arendt, *Between Past and Future: Six Exercises in Political Thought* (Cleveland, N.Y.: Meridian Books, 1963), p. 148.

42. The idea that Heidegger is creating a space for anarchy comes from Reiner Schurmann's *Heidegger on Being and Acting*. Schurmann's book has deeply affected my reading of Heidegger. His thought is behind my translation of Enframing as "Technoarchy."

43. Heidegger, "On The Essence of Truth," p. 138.

44. Dallmayr, "Ontology of Freedom," p. 204.

45. Jack Weatherford, *Indian Givers: How the Indians of the Americas Transformed the World* (New York: Fawcett Columbine, 1988), p. 145.

46. William Bevis, *Ten Tough Trips: Montana Writers and the West* (Seattle: University of Washington Press, 1990), pp. 111–12.

47. Max Oelschlaeger, *The Idea of Wilderness* (New Haven, Conn.: Yale University Press, 1991), p. 275.

3. Science and Technology

1. Heidegger, "The Question concerning Technology," p. 22.

2. Robert Ackermann: *Data, Instruments, and Theory: A Dialectical Approach to Understanding Science* (Princeton, N.J.: Princeton University Press, 1985), p. 125.

3. Martin Heidegger, "The Age of the World Picture," in *The Question concerning Technology*, trans. William Lovitt (New York: Harper and Row, 1977), p. 122.

4. Martin Heidegger, "Science and Reflection," in *The Question concerning Technology*, p. 158.

5. Nicolas Platon, *Crete* (Geneva: Nagel Publishers, 1966), p. 148.

6. Heidegger, "The Age of the World Picture," p. 123.

7. Ibid., p. 121.

8. Paul Feyerabend, *Against Method* (London: Verso Editions, 1980), p. 97.

9. Heidegger, "The Age of the World Picture," p. 121.

10. For a vigorous defense of Aristotle's science, see "Aristotle Not a Dead Dog," in Paul Feyerabend's *Science in a Free Society* (London: Verso Editions, 1978).

11. William Barrett, *The Illusion of Technique* (London: William Kimber, 1979), p. 89.

12. Douglas R. Hofstadter, *Gödel, Escher, Bach: An Eternal Golden Braid* (New York: Random House, 1980), p. 228.
13. Heidegger, "Science and Reflection," p. 167.
14. Heidegger, "The Age of the World Picture," p. 133.
15. Ibid., p. 120.
16. Ibid., p. 121.
17. Ibid., p. 125.
18. Ibid., p. 126.
19. Jerome Ravetz, *Scientific Knowledge and Its Social Problems* (New York: Oxford University Press, 1971), p. 273.
20. "Methodism" can be a pretty demanding religion, needing many bloody sacrifices. See Donald McCloskey, *The Rhetoric of Economics* (Madison: University of Wisconsin Press, 1985), p. 11.
21. Ackermann, *Data, Instruments, and Theory*, p. 35.
22. Heidegger, "The Age of the World Picture," p. 125.
23. Quoted from D. M. Dooling, "Focus," *Parabola* 16, no. 3 (August 1991): 3.
24. Heidegger, "The Age of the World Picture," p. 127.
25. Ibid., p. 129.
26. Ibid., p. 126 and appendix 4.
27. Ibid., p. 128.
28. Ibid., p. 129.
29. I read about this incident in several science magazines several years ago. I cannot find the source now.
30. Heidegger, "The Age of the World Picture," p. 129.
31. Ibid., p. 141.
32. Ibid., p. 130.
33. Ibid., p. 132.
34. Michel Foucault makes the same argument. See his *Order of Things*, trans. Alan Sheridan (New York: Vintage Books, 1973), p. 312.
35. Heidegger, "The Age of the World Picture," p. 132.
36. Foucault, *The Order of Things*, p. 348.
37. Ibid., p. 344.
38. Heidegger, "The Age of the World Picture," p. 133.
39. See my unpublished essay "Toleration and Shamanism."

4. Technoarchy

1. James R. Beniger, *The Control Revolution: Technological and Economic Origins of the Information Society* (Cambridge: Harvard University Press, 1986), pp. 433–34.
2. Brown, *Bury My Heart at Wounded Knee.*

3. T. C. McLuhan, comp., *Touch the Earth* (New York: Simon and Schuster, Touchstone Books, 1971), p. 15.

4. Heidegger, "The Question concerning Technology," p. 15.

5. Wendell Berry, *The Unsettling of America* (San Francisco: Sierra Club Books, 1977), p. 17.

6. Heidegger, "The Question concerning Technology," p. 16.

7. Postmodernism can be read as an attempt to escape somehow the imperatives of technology; see Henry Kariel, *The Desperate Politics of Postmodernism* (Amherst: University of Massachusetts Press, 1989), p. 10.

8. Heidegger, "The Question concerning Technology," p. 18.

9. Ibid., p. 16.

10. Ibid., p. 17.

11. Ibid., p. 18.

12. Ibid., p. 27.

13. John Kenneth Galbraith, *Economics and Public Purpose* (New York: New American Library, 1973), p. 405.

14. Frank Oski, "Formula for Profits: Heating Up the Bottle Battle," *Nation* 249, no. 19 (December 4, 1989): 665.

15. This joke became the subject of much media controversy during the farm depression in 1986.

16. Heidegger, "The Question concerning Technology," p. 18.

17. Ibid., p. 19.

18. Ibid., p. 20.

19. Ibid., p. 21.

20. Barrett, *The Illusion of Technique*, p. 187.

21. Heidegger, "The Question concerning Technology," p. 21.

22. Ibid., p. 25.

23. My reading of anarchy comes from Schurmann, *Heidegger on Being and Acting*, p. 25.

24. Heidegger, "The Question concerning Technology," p. 26.

25. Falwell often boasts on his TV program of how all the teachers at his Bible college swear oaths that they believe in Creationism.

26. Heidegger, "The Question concerning Technology," p. 27.

27. Ibid., p. 28.

28. Ibid., p. 31.

5. The Flight of the Gods

1. For a discussion of destiny, see Martin Heidegger, "The Turning," in *The Question concerning Technology*, trans. William Lovitt (New York: Harper and Row, 1977), p. 38.

2. Nietzsche writes: "For some time now, our whole European culture has been moving as toward a catastrophe, with a tortured tension that is growing from decade to decade: restlessly, violently, headlong, like a river that wants to reach the end, that no longer reflects, that is afraid to reflect" (*The Will to Power*, ed. Walter Kaufmann [New York: Vintage Books, 1974], p. 3).

3. Stuart Hampshire, *Freedom of the Individual* (Princeton, N.J.: Princeton University Press, 1977).

4. On the confession of truth, see Michel Foucault, *The History of Sexuality*, vol. 1, trans. Robert Hurley (New York: Random House, 1973), p. 17.

5. For a discussion of the three "masters of suspicion," see Paul Ricoeur's *Freud and Philosophy: An Essay on Interpretation* (New Haven, Conn.: Yale University Press, 1970), p. 32.

6. Foucault, *The Order of Things*, p. 328.

7. Bill Connolly argued in a letter to me that Nietzsche is not a technocrat, and I agree. Nietzsche is not an unambiguous supporter of modern science and technology, but rather an interesting critic of them. One of Nietzsche's most haunting images is that of the man of science on his belly, groveling before the facts. This is certainly not the attitude of an overman! Science, technology, and the machine, according to Nietzsche in the *Will to Power*, have leveled out all of humanity, submerging them all in the purposeless necessity of the mechanism. With no meaning to govern their actions, reduced uniformly to the machine's utility and bred as a herd, the way is opened for the overman, the one who will use the herd to create his own meaning, impose his will on everything. On the one hand, modern science and technology lead toward the ultimate sickness of the will, the breeding of the herd; on the other, the herd becomes the ultimate means for the overman, an instrument for his use, for him to affirm his essence.

Even though Nietzsche, as always, is ambiguous in his evaluation of science and technology, and even though he writes of self-overcoming, he nevertheless continues, as all of modernity does, to think of science and technology as a means available for the will, ultimately the overman's will—not as a way, limited by its time and place, of bringing forth truth. Taking the will to power as the truth of all beings, appropriating everything as a means to the will's willing, Nietzsche does not overcome the age of the machine but brings it to its highest fulfillment, the great noon of its triumph. By bringing everything forth as a means for the will to power, Nietzsche thinks the most appropriate thought of our time, the pure utility of everything. In so doing, he opens himself up to the most horrible question of all times—What is anything for? If his answer—the overman—does not overcome the age of the machine, it does enable us (we who interpret the thought of the last metaphysician) to think the thought that would at last reveal the mystery of Being.

Nietzsche existed; his thought is not an error to be overcome but a truth too terrible to ignore. In his thought, the flight of the gods becomes an absence that we must at last acknowledge, and because of this, they become present as an absence. Now we must ask of everything, What for? And in our use of things and people, we must acknowledge in a new way that we have no answer. Nietzsche does that to us.

If we think Nietzsche's truth as the abyss of Being opening up in our presence, we can know his nihilism as something divine, something the holy sent to us. Nietzsche himself recognized the divinity of his nihilism when he cried, "I seek God, I seek God." Nietzsche was a poet, and we should think his thought as poetry, acknowledging its ambiguity and multiple meanings while we listen to the whisper of the world worlding through it.

8. Nietzsche, *Will to Power*, p. 10.

9. Ibid., p. 24.

10. Nietzsche, "The Madman," in *The Gay Science*, in Walter Kaufmann's *Portable Nietzsche* (New York: Viking, 1968), pp. 95–96.

11. Martin Heidegger, "The Word of Nietzsche," in *The Question concerning Technology*, trans. William Lovitt (New York: Harper and Row, 1977), p. 61.

12. Foucault, *The Order of Things*, p. 312.

13. Heidegger, "The Word of Nietzsche," p. 62.

14. Ibid., p. 64.

15. Ibid., p. 98.

16. Stanley Rosen describes the discontinuity when tradition comes under attack, in his *Nihilism: A Philosophical Essay* (New Haven, Conn.: Yale University Press, 1969), p. 230.

17. Heidegger, "The Word of Nietzsche," p. 66.

18. Ibid., p. 61.

19. For a discussion of metaphysics, see Martin Heidegger's *Introduction to Metaphysics*, trans. Ralph Manhiem (New Haven, Conn.: Yale University Press, 1976).

20. Michael Allen Gillespie, *Hegel, Heidegger, and the Ground of History* (Chicago: University of Chicago Press, 1984), p. 146.

21. He certainly does like scandalizing Christian morality. See Wendell Berry, "The Mad Farmer Revolution," in *Farming: A Hand Book* (New York: Harcourt Brace Jovanovich, 1970), p. 42. His notion of resurrection is distinctly Greek, more cyclic than apocalyptic.

22. Heidegger, "The Word of Nietzsche," p. 64.

23. Reinhold Niebuhr offers a Christian reading of Nietzsche that is interesting; see his *Beyond Tragedy* (New York: Charles Scribner's Sons, 1937), p. 195.

24. The "existence" of "God" may be necessary for us to communicate with each other. In Kafka's novels people are continually misunderstanding each other—they have lost the center which would make it possible for them to read each other. See Peter Heller, *Dialectics and Nihilism: Essays on Lessing, Nietzsche, Mann, and Kafka* (Amherst: University of Massachusetts Press, 1966), p. 250.

25. Heidegger, "The Word of Nietzsche," p. 54.

26. Ibid., p. 84.

27. Ibid., p. 99.

28. Ibid., p. 100.

29. Ibid., p. 102.

30. Ibid., p. 103.

31. David Michael Levin raises the problem of the concealment of Being in terms of "vision" in his *Opening of Vision: Nihilism and the Postmodern Situation* (New York: Routledge, 1988), p. 152.

32. Heidegger, "The Word of Nietzsche," p. 104.

33. Ibid., p. 105.

34. Gilles Deleuze offers a different reading of nihilism, as being the production of reactive mentality; see his *Nietzsche and Philosophy*, trans. Hugh Tomlinson (New York: Columbia University Press, 1983), p. 147.

35. Heidegger, "The Word of Nietzsche," p. 105.

36. Ibid., p. 108.

37. Ibid., p. 109.

39. Ibid., p. 110.

6. A Prison of Freedom

1. For an alternative understanding of power, see Michel Foucault, "Two Lectures," in *Power-Knowledge*, trans. Colin Gordon, Leo Marshall, John Mepham, and Kate Soper (New York: Pantheon Books, 1977), p. 78.

2. For a discussion of the underlying unity of Americanism, Naz-

ism, and Marxism, see Gillespie, *Hegel, Heidegger, and the Ground of History*, p. 131.

3. Foucault argues that it is not the individual that is the agent but institutions, discursive practices, rituals, and so on ("Two Lectures," p. 101).

4. Foucault defines power nicely in *The History of Sexuality*, vol. 1, p. 92. Heidegger is more circumspect; see "The Question concerning Technology," p. 25.

5. Michel Foucault writes in many places about how reason isolates its others. One is *Madness and Civilization*, p. 221.

6. Michel Foucault, *Discipline and Punish*, trans. Alan Sheridan (New York: Pantheon Books, 1977), p. 30.

7. Ibid., p. 28.

8. Ibid., p. 135.

9. Tom Keenan, "The 'Paradox' of Knowledge and Power: Reading Foucault on a Bias," *Political Theory* 15, no. 1 (February 1987): 5.

10. Heidegger, "The Word of Nietzsche," p. 79.

11. For a discussion of Foucault's "placeless place" in discourse, see Tom Keenan, "The 'Paradox' of Knowledge and Power," p. 14.

12. This rupture, like all of Foucault's breaks, happens without identifiable cause. History has no teleology under which such a cause could be understood (*Discipline and Punish*, p. 57).

13. Lewis Mumford, *Technics and Civilization* (New York: Harcourt Brace Jovanovich, 1962), p. 124.

14. Ibid., p. 125.

15. Ibid., p. 128.

16. Foucault, *Discipline and Punish*, p. 104.

17. Ibid., p. 106.

18. This reading of the social sciences is from Hubert Dreyfus and Paul Rabinow's book *Michel Foucault: Beyond Structuralism and Hermeneutics* (Chicago: University of Chicago Press, 1982), p. 134.

19. Foucault, *Discipline and Punish*, p. 109.

20. Ibid., p. 113.

21. Albert O. Hirschman, *The Passions and the Interests: Political Arguments for Capitalism before Its Triumph* (Princeton, N.J.: Princeton University Press, 1977), p. 15.

22. Ibid., p. 18.

23. Ibid., p. 12.

24. Ibid., p. 53.

25. William Connolly, *Political Theory and Modernity* (New York: Basil Blackweil, 1988), p. 28.

26. Richard Hofstadter, "The Founding Fathers: An Age of Realism," in *The Moral Foundations of the American Republic*, ed. Robert Horwitz (Charlottesville: University Press of Virginia, 1979), p. 77.
27. Martin Diamond, "Ethics and Politics: The American Way," in *The Moral Foundations of the American Republic*, ed. Robert Horwitz (Charlottesville: University Press of Virginia, 1979), p. 48.
28. Quoted in Robert N. Bellah, Richard Madsen, William M. Sullivan, Ann Swidler, and Steven M. Tipton, *Habits of the Heart: Individualism and Commitment in American Life* (Berkeley: University of California Press, 1985), p. 28.
29. Ibid., p. 29.
30. Quoted from ibid., p. 31.
31. Ibid., p. 255.
32. Hirschman, *The Passions and the Interests*, p. 10.
33. Ibid., p. 17.
34. Ibid., p. 28.
35. Ibid., p. 32.
36. Ibid., p. 44.
37. Ibid., p. 79.
38. Foucault brilliantly describes the nature of the medieval world (*The Order of Things*, p. 17).
39. Thomas Hobbes, *Leviathan*, parts 1, 2 (Indianapolis: Bobbs-Merrill, 1958), p. 140.
40. Foucault, *Discipline and Punish*, p. 123.
41. Thomas L. Dumm, *Democracy and Punishment: Disciplinary Origins of the United States* (Madison: University of Wisconsin Press, 1987), p. 95.
42. Quoted in ibid., p. 92.
43. Foucault, *Discipline and Punish*, p. 136.
44. Ibid., p. 141.
45. Ibid., p. 149.
46. Ibid., p. 170.
47. Ibid., p. 141.
48. Ibid., p. 145.
49. Ibid., p. 184.
50. Ibid., p. 171.
51. Ibid., p. 154.
52. Ibid., p. 172.
53. Ibid., p. 175.
54. Ibid., p. 177.
55. Ibid., p. 184.

56. Ibid., p. 189.
57. Ibid., p. 195.
58. Ibid., p. 200.
59. Ibid., p. 202.
60. Ibid., p. 203.
61. Hirschman, *The Passions and the Interests*, p. 30.
62. Madison writes: "In framing a government which is to be administered by men over men, the great difficulty lies in this: you must first enable the government to control the governed; and in the next place oblige it to control itself." See Alexander Hamilton, James Madison, and John Jay, *The Federalist Papers* (New York: New American Library, 1961), p. 322.
63. Dumm, *Democracy and Punishment*, p. 33.
64. James Bamford, *The Puzzle Palace: A Report on NSA, America's Most Secret Agency* (Boston: Houghton Mifflin, 1982), p. 305.
65. Foucault, *Discipline and Punish*, p. 276.

7. The Collapse of the Household

1. Berry, *The Unsettling of America*, p. 11.
2. Neil R. Sampson, *Farmland or Wasteland: A Time to Choose* (Emmaus, Pa.: Rodale Press, 1981), p. 10. Lester Brown makes the same point, but with better numbers, in *State of the World, 1989* (New York: W. W. Norton, 1989), p. 43.
3. Heidegger, "The Question concerning Technology," p. 17.
4. Ruth Schwartz Cowan, *More Work for Mother* (New York: Basic Books, 1983), p. 19.
5. For a historical example of a "partnership" between the sexes, see Eisler, *The Chalice and the Blade*, chaps. 2 and 3.
6. Cowan, *More Work for Mother*, chap. 4.
7. Best and Connolly, *The Politicized Economy*, pp. 54–55.
8. Ibid., chap. 3.
9. Cowan, *More Work for Mother*, p. 63.
10. Ibid., p. 51.
11. Ibid., chap. 6.
12. Ibid., p. 48.
13. Berry, *The Unsettling of America*, chap. 1.
14. U.S. Bureau of the Census, *Statistical Abstract of the United States*, p. 649.
15. Jean-Pierre Berlan and Richard Lewontin, "Technology, Re-

search, and the Penetration of Capital: The Case of U.S. Agriculture," *Monthly Review* (July–August 1986), p. 26.

16. Ibid., p. 27.

17. Ibid., p. 28.

18. Ibid., p. 27.

19. Ibid., p. 34.

20. Cowan, *More Work for Mother*, p. 63.

21. Lillian Breslow Rubin, *Worlds of Pain: Life in the Working-Class Family* (New York: Basic Books, 1977), p. 164.

22. Ibid., p. 159.

23. Ibid., p. 113.

24. Ibid., p. 169.

25. Berry, *The Unsettling of America*, pp. 117–40.

26. I remember this incident from a "Sixty Minutes" program on migrant workers.

8. Harnessing the Earth to the Slavery of Man

1. Marx gets carried away with himself, for example, in the "Communist Manifesto," where he sings hymns of praise to the machine (in *The Marx-Engels Reader*, ed. Robert C. Tucker [New York: W. W. Norton, 1978]).

2. Lenin speaks glowingly of Taylor's "scientific management" of the work force, in "The Immediate Tasks of the Soviet Government" (1918), in *Collected Works*, vol. 27 (Moscow: International Press, 1965), p. 259.

3. This definition of the machine is a reading of Lewis Mumford's definition (*Technics and Civilization*, p. 9).

4. Machines can be thought of as information systems. See Jeremy Campbell, *Grammatical Man: Information, Entropy, Language, and Life* (New York: Simon and Schuster, 1982), p. 104.

5. Karl Marx, *Capital*, vol. 1 (New York: International Publishers, 1977), pp. 353–54.

6. Michael Best pointed out to me in a letter that the degradation of work in a capitalist economy is not always as smooth and continuous as Braverman argues, but rather that a complex dialectic takes place between the machine and the worker. For instance, the machine may so completely replace the worker that there is nowhere for him or her to go except to a more highly skilled job. Or the development of machine technology may be hampered by the ready availability of cheap workers. It seems that the introduction of the

machine and the degradation of the worker take place in fits and starts. The process sometimes reverses itself for a while, before conditions are appropriate for more rationalization of production. After years of having their work increasingly rationalized, blue-collar workers in some industries, usually where production proceeds by batch process instead of a continuous assembly line, are finding their skills increasing. However, many low-level white-collar workers, traditionally spared from the rigors of rationalized work, are finding their work increasingly disciplined. No doubt the way is being opened up for their replacement by artificial intelligence programs. And as I observed, many highly skilled jobs have been opening up in areas such as advertising, security, labor management, and correction.

Despite this flux, the essential degradation of humanity continues on because from the very beginning we have already been appropriated in our entirety as utility for the machine. It does not matter if some find their skill level increasing, because it is only because the rationality of production at the moment dictates it. As long as humanity is as utility for production, the skill of human beings will be subjected to the measure of efficiency, the rationality of production, and the utility of systemic demand. Appropriated entirely by reason, it will be a skill for bringing about the distant imperatives of Technoarchy, not a skill for bringing the thing forth from the earth in response to the call of the dwelling place.

7. Frederick W. Taylor, *The Principles of Scientific Management* (New York: W. W. Norton, 1967), p. 36.

8. Ibid., pp. 37–38.

9. Ibid., p. 63.

10. William R. Spriegel and Clark E. Myers, eds., *The Writings of the Gilbreths* (Homewood, Ill.: R. D. Irwin, 1953).

11. Harry Braverman, *Labor and Monopoly Capital: The Degradation of Work in the Twentieth Century* (New York: Monthly Review Press, 1974), p. 177.

12. Ibid., p. 188.

13. Ibid., p. 189.

14. Ibid., p. 190.

15. The classic on cybernetics is Ludwig von Bertalanffy's *General System Theory* (New York: George Braziller, 1968), pp. 44, 78, 90, 150, 161.

16. Braverman, *Labor and Monopoly Capital*, p. 192.

17. Lewis Mumford, *The Pentagon of Power: The Myth of the Machine* (New York: Harcourt Brace Jovanovich, 1970), p. 241.

18. James Bright, "The Relationship of Increasing Automation and Skill Requirements," in National Commission on Technology, Automation, and Economic Progress, *The Employment Impact of Technological Change*, Appendix, vol. 2, *Technology and the American Economy* (Washington, D.C.: U.S. Government Printing Office, 1966), p. 220.

19. James Bright, *Automation and Management* (Boston: Maxwell, 1958), pp. 218–19.

20. Rubin, *Worlds of Pain*, pp. 159–60.

21. Ibid., pp. 160–61.

22. Berry, *The Unsettling of America*, p. 93.

23. Ibid., p. 87.

24. Ibid., p. 91.

25. Ibid., p. 90.

26. Ibid., p. 93.

27. I heard these astonishing facts over the radio as I was writing this chapter. A government study by the Office of Technology Assessment is the origin.

28. Karl Marx, *Capital*, vol. 1; see "Primitive Accumulation."

29. Best and Connolly, *The Politicized Economy*, p. 19.

30. On the creation of a rabble disaffected from the economy, see G. W. F. Hegel, *Phenomenology of the Spirit*, trans. A. V. Miller (Oxford: Oxford University Press, 1979), para. 73 on p. 46.

31. Heidegger has reflected deeply on the meaning of death. See James M. Demske, *Being, Man, and Death: A Key to Heidegger* (Lexington: University Press of Kentucky, 1970).

9. The Vulnerable Machine

1. *Time*, March 3, 1986, "Periscope."

2. Many examples of postmodernism can be given, but an especially good one is Mark C. Taylor, *Erring: A Postmodern A-theology* (Chicago: University of Chicago Press, 1984).

3. Related to me by my uncle, Mike Sikorski, a retired air force major.

4. Neil Sheehan, *A Bright Shining Lie: John Paul Vann and America in Vietnam* (New York: Random House, 1988), pp. 675–81.

5. Bob Woodward, *The Commanders* (New York: Simon and Schuster, 1991), p. 376.

6. Ibid., p. 330.

7. Blair, *Strategic Command and Control*, appendixes C and D.

8. Astonishingly, none of our nuclear power plants was built with any EMP protection; see Amory B. Lovins and Hunter Lovins, *Brittle Power: Energy Strategy for National Security* (Andover, Mass.: Brick House, 1982), p. 73.

9. Daniel Ford, *Meltdown: The Secret Papers of the Atomic Energy Commission* (New York: Simon and Schuster, Touchstone Books, 1986), pp. 95–98.

10. Lovins and Lovins, *Brittle Power*, p. 177.

11. Ibid., p. 59.

12. Ibid., p. 60.

13. Ibid., p. 62.

14. Ibid., p. 218.

15. Ibid., p. 177.

16. Ibid., p. 62.

17. Ibid., p. 182.

18. John Kenneth Galbraith makes the argument that consumers are not the sovereigns of the marketplace; rather, monopolies are (*The New Industrial State* [New York: New American Library, 1985], p. 194).

19. This general argument is made by Roberto Vacca, *The Coming Dark Age* (New York: Anchor Books, 1974), as well as by Alvin Toffler, *The Eco-Spasm Report* (New York: Bantam Books, 1975).

20. On the incompleteness of systems, see Hofstadter, *Gödel, Escher, Bach.* On the nature of chaos, see James Gleick, *Chaos: Making a New Science* (New York: Penguin Books, 1987).

21. Best and Connolly, *The Politicized Economy*, p. 155.

22. Charles L. Schultze, *The Public Use of Private Interest* (Washington, D.C.: Brookings Institution, 1977), p. 31.

23. Ibid., p. 49.

24. William Connolly, *Appearance and Reality in Politics* (Cambridge: Cambridge University Press, 1981), p. 104.

25. For a discussion of this dynamic, see Michel Serres, *The Parasite*, trans. Lawrence R. Schehr (Baltimore: Johns Hopkins University Press, 1982).

26. Berry, *The Unsettling of America*, p. 70.

27. Lovins and Lovins, *Brittle Power*, p. 22.

10. The Monster

1. In 1810, the count de Montalivet, minister of the interior, issued this order: “Considering that M. de Sade is suffering from the

most dangerous of insanities, contact between him and the other inmates poses incalculable dangers, and for as much as his writings are no less demented than his speech and conduct, I therefore order the following: That Monsieur de Sade be given completely separate lodging so that he be barred from all communication with others, and that the greatest care be taken to prevent any use by him of pencils, pens, ink, or paper" (Sade, *The Marquis de Sade*, trans. Richard Seaver and Austryn Wainhouse [New York: Grove Press, 1965], p. 116).

2. Angela Carter, *The Sadeian Woman and the Ideology of Pornography* (New York: Pantheon Books, 1978), p. 32.

3. Sade's relation to reason is ambiguous. See Foucault, *Madness and Civilization*, p. 285.

4. On the confession of perversion, see Foucault, *The History of Sexuality*, vol. 1, p. 21.

5. Ibid., p. 150.

6. On the problems that the double poses for a theory of legitimacy, see William Connolly, *Politics and Ambiguity* (Madison: University of Wisconsin Press, 1987), p. 89.

7. Foucault, *Madness and Civilization*, p. 20.

8. Ibid.; see "The New Division."

9. Heidegger, "The Age of the World Picture," p. 127.

10. Mary Shelley, *Frankenstein; or, The Modern Prometheus* (New York: New American Library, 1965).

11. Barbara Johnson, "Le dernier homme," in *Les fins de l'homme: A partir du travail de Jacques Derrida*, ed. Philippe Lacoue-Labarthe and Jean-Luc Nancy (Paris: Editions Galilée, 1980), p. 77.

12. Shelley, *Frankenstein*, p. 53.

13. Ibid., p. 56.

14. Ibid., p. 57.

15. Ibid., p. 87.

16. Ibid., p. 138.

17. Ibid., p. 159.

18. Jean-Jacques Rousseau, *First and Second Discourses*, ed. Roger Masters (New York: St. Martin's Press, 1964), p. 149.

19. Jean-Jacques Rousseau, *On the Social Contract: With Geneva Manuscript and Political Economy*, ed. Roger Masters, trans. Judith Masters (New York: St. Martin's Press, 1978), p. 56.

20. Shelley, *Frankenstein*, p. 115.

21. Ibid., p. 139.

22. Ibid., p. 160.

23. Sex is not the purpose of Sade's writings, but it is his strategy for expressing power and dominance. See Carter, *The Sadeian Woman and the Ideology of Pornography*, p. 24.

24. The Marquis de Sade, "Philosophy in the Bedroom," in *The Marquis de Sade*, trans. Richard Seaver and Austryn Wainhouse (New York: Grove Press, 1965), p. 238.

25. Connolly argues that Sade and Rousseau are doubles (*Political Theory and Modernity*, p. 74).

26. I do not want to give the impression in this chapter that Rousseau is a secularist and that Sade shows him what his secularism comes to. As William Connolly has reminded me in a letter, God is very important to Rousseau, the ultimate defense against the monster Sade. If there is no God, then there is no design, purpose, or end to anything and everything is permitted, even the most horrible abominations. Rousseau understands Sade very well because Sade is the abyss that he would fall into if it were not for God. Connolly argues, and I agree, that Rousseau finds the voice of God in nature, that he follows the Enlightenment in the thought that God can no longer be known in scripture, through miracles, or in the authority of the church. He is revealed in nature because nature is a coherence of forces, each linked to another, efficiently transmitting his will as the first mover throughout the whole extent of the universe. Because nothing material can of itself move but only transmit the motion it receives from another body, the chain of cause and effect throughout the universe must eventually lead back to a nonmaterial will that wills everything and first sets it into motion—God. That everything in the universe fits together so nicely, so perfectly reflecting a rational design, is a testament to the intelligence of the being that set everything into motion. As a coherence of cause and effect, nature is thus the proof and reflection of a rational will, a meaningful design, and a supernatural purpose.

For Rousseau, however, nature itself is nothing holy but is only the material manifestation and proof of a nonmaterial will. God's design and will can be discerned in nature, known through it; nature, however, being material and linked together by a chain of cause and effect, is something radically different from God. Rousseau, I believe, follows the Enlightenment and the history of Christianity in general by removing God from nature to a beyond, a supernatural realm where the divine will designs nature according to its intelligence and sets it into motion according to its will. Even though Rousseau can hear the voice of God through nature, he has radically purged God from it by making God something wholly other than it, a nonmaterial will that expresses its design through the linkage of cause and effect. Thus, even though God is the ultimate source of Rousseau's order, the ways of nature that reveal God's will are, of themselves, nothing holy. In this, Rousseau has the same relationship to nature that Sade does. It is just without the design and meaning of a supernatural will. This is why Rousseau and the Enlightenment are so haunted by Sade. Since the Enlightenment continues what Christianity began and removes God from nature to a supernatural beyond, far from the earth in a will that wills everything at the beginning of time, it is but a short step to declaring that there is no God and thus that nature has no plan, meaning, or purpose. Sade de-secrates nature, but only because nature has already been desecrated by the retreat of God into the supernatural.

And yet Rousseau finds in nature the truth which is the ground of the legitimate state. God does not reveal his will in scripture, at least not unambiguously and without the corruption of Man entered into it, but it can be discerned in nature, de-secrated though it is, because it is the occurrence of God's will. From the nature of Man, Man can discern the will of God. And as a will that wills its nature, Man can discern in his willing God's will for Man. But—and this is crucial—we do not need to refer to God, except implicitly, to establish the ground of the legitimate state because everything necessary for that is contained in Man's nature and his will. Rousseau never refers to the supernatural as the ground of the legitimate state in his social contract, only to the natural. And so, although God is crucial to Rousseau's metaphysic, he has removed himself from Man's natural existence. With Rousseau, and the Enlightenment in general, Man is liberated from any need to refer to the supernatural as he goes about his natural affairs. And so, even those who profess strong belief in God become political secularists. This is what is meant by the phrase "the flight of the gods."

27. Rousseau, *On the Social Contract*, p. 49.

28. Martin Heidegger, "The Word of Nietzsche: God Is Dead," in *The Question concerning Technology*, trans. William Lovitt (New York: Harper and Row, 1977), p. 87.

29. Rousseau, *On the Social Contract*, p. 61.

30. Ibid., p. 71.

31. Ibid., p. 55.

32. Connolly, *Political Theory and Modernity*, p. 72.

33. Sade, "Philosophy in the Bedroom," pp. 208–9.

34. Maurice Blanchot, "Sade," in *The Marquis de Sade*, trans. Richard Seaver and Austryn Wainhouse (New York: Grove Press, 1965), p. 52.

35. Hegel, *The Phenomenology of Spirit*, para. 188 on p. 115.

36. The Marquis de Sade, "Justine," in *The Marquis de Sade* (New York: Grove Press, 1965), p. 492.

37. Sade, "Philosophy in the Bedroom," p. 337.

11. The Turning

1. Michael E. Zimmerman, "The Thorn in Heidegger's Side: The Question of National Socialism," *Philosophical Forum* 20, no. 4 (Summer 1989): 351.

2. They write: "Farias himself was very nearly unknown until the book's appearance. He is obviously not the most expert chronicler. . . . As an outsider, something of a marginal figure, he was not really susceptible to the usual professional prudence or 'correction' " (Tom Rockmore and Joseph Margolis, foreword to *Heidegger and Nazism*, by Victor Farias [Philadelphia: Temple University Press, 1989], p. xx).

3. Hans-Georg Gadamer, "Back from Syracuse?" *Critical Inquiry* 15 (Winter 1989): 429.

4. All word etymologies are from the *Oxford English Dictionary*.

5. M. Scott Peck, *People of the Lie* (New York: Simon and Schuster, Touchstone Books, 1983), p. 69.

6. Joseph Campbell, with Bill Moyers, *The Power of Myth* (New York: Doubleday, 1988), p. 48.

7. Elaine Pagels, *Adam, Eve, and the Serpent* (New York: Random House, 1988), p. 71.

8. As an example of this relationship, see Heinrich Kramer and James Sprenger, *The Malleus Maleficarum* (New York: Dover Publications, 1971). Written in the Middle Ages as a guide for witch-hunters, it was supposedly intended for good, yet few books can be judged so evil in its effects.

9. Jacques Derrida, *Of Spirit: Heidegger and the Question*, trans. Geoffrey Bennington and Rachel Bowlby (Chicago: University of Chicago Press, 1989), p. 121.

10. Derrida himself started the controversy off with an interview in *Le Nouvel Observateur*, November 6–12, 1987, called "Heidegger, l'enfer des philosophes."

11. Derrida, *Of Spirit*, p. 2.

12. Heidegger writes: "The object-character of technological dominion spreads itself over the earth ever more quickly, ruthlessly, and completely. Not only does it establish all things as producible in the process of production; it also delivers the products of production by means of the market. In self-assertive production, the humanness of man and the thingness of things dissolve into the calculated market value of a market which not only spans the whole earth as a world market, but also, as the will to will, trades in the nature of Being and thus subjects all beings to the trade of a calculation that dominates most tenaciously in those areas where there is no need of numbers" ("What Are Poets For?" in *Poetry, Language, Thought*, trans. Albert Hofstadter [New York: Harper and Row, 1971], p. 114).

13. Heidegger, "Science and Reflection."

14. Martin Heidegger, "The Self-Assertion of the German University: Address Delivered on the Solemn Assumption of the Rectorate of the University of Freiburg" and "The Rectorate 1933/34: Facts and Thoughts," trans. Karsten Harries, *Review of Metaphysics* 38 (March 1985): 467–502.

15. Ibid., pp. 479–80.

16. Arnold I. Davidson, "Symposium on Heidegger and Nazism," *Critical Inquiry* 15 (Winter 1989): 417.

17. Ibid., p. 419.

18. Heidegger, "The Turning," p. 41.

19. Ibid., p. 43.

20. Ibid., p. 41.

21. Ibid., p. 39.

22. Michel Foucault, "Man and His Doubles," in *The Order of Things*, trans. Alan Sheridan (New York: Random House, 1973).

23. Foucault, *Discipline and Punish*, p. 30.

24. Heidegger writes: "Correspondingly, human willing too can be in the mode of self-assertion only by forcing everything under its dominion from the start, even before it can survey it. To such a willing, everything, beforehand and thus subsequently, turns irresistibly into material for self-assertive production. The earth and its atmosphere become raw material. Man becomes human material, which is disposed of with a view to proposed goals. The unconditioned establishment of the unconditional self-assertion by which the world is purposefully made over according to the frame of mind of man's command is a process that emerges from the hidden nature of technology" ("What Are Poets For?" p. 111).

25. Quoted by Philippe Lacoue-Labarthe in *Heidegger, Art, and*

Politics, trans. Chris Turner (Cambridge: Basil Blackwell, 1990), p. 34. His source is Wolfgang Schirmacher, *Technik und Gelassenheit* (Freiburg: Karl Alber, 1984).

26. For a discussion of the similarity between the two camps, see Milton Mayer, *They Thought They Were Free* (Chicago: University of Chicago Press, 1955), chap. 1.

27. Lacoue-Labarthe, *Heidegger, Art, and Politics*, p. 37.

28. Brown, *State of the World, 1987*, p. 122.

29. Heidegger, "The Turning," p. 39.

30. Ibid., p. 43.

31. Ibid., p. 38.

32. Ibid., p. 40.

33. Ibid., p. 39.

34. Schurmann, *Heidegger on Being and Acting*, p. 72.

35. Heidegger, "The Turning," p. 45.

36. Ibid., p. 48.

37. Reiner Schurmann brilliantly explores the possibilities of anarchy in *Heidegger on Being and Acting*, p. 275.

38. Gillespie, *Hegel, Heidegger, and the Ground of History*, p. 174.

39. Ibid.

40. Schurmann, *Heidegger on Being and Acting*, p. 235.

41. Heidegger, "The Turning," p. 42.

42. Ibid., p. 46.

12. Building Wilderness

1. I got the idea of the ark from the *Journal of the New Alchemists*, issue no. 7 in particular (Brattleboro, Vt.: Stephen Greene Press, 1981).

2. Henry David Thoreau, *Walden and Other Writings* (New York: Bantam Books, 1962), p. 172.

3. Any readers wanting advice on building their own house may contact me at home, Box 202, Willard, MT 59354, telephone 406-775-6378. Some do-it-yourself books that I used and can recommend: Mario Salvadori, *Why Buildings Stand Up: The Strength of Architecture* (New York: W. W. Norton, 1980); Bill Keisling, *Solar Water Heating Systems* (Emmaus, Pa.: Rodale Press, 1983); Edward Mazria, *The Passive Solar Energy Book* (Emmaus, Pa.: Rodale Press, 1979); Robert L. Roy, *Underground Houses* (New York: Sterling Publishing, 1982); David Martindale, *Earth Shelters* (New York: E. P. Dutton, 1981); Underground Space Center, University of Minnesota,

Earth Sheltered Housing Design: Guidelines, Examples, and References (New York: Van Nostrand Reinhold, 1978); and Alex Wade and Neal Ewenstein, *Thirty Energy-Efficient Houses You Can Build* (Emmaus, Pa.: Rodale Press, 1977).

4. Schurmann, *Heidegger on Being and Acting*, p. 97.
5. Martin Heidegger, "Building Dwelling Thinking," in *Poetry, Language, Thought* (New York: Harper and Row, 1971), p. 147.
6. Berry, *The Unsettling of America*, p. 87.
7. Heidegger, "Building Dwelling Thinking," p. 147.
8. Heidegger, "What Are Poets For?" p. 111.
9. Heidegger, "Building Dwelling Thinking," p. 148.
10. Heidegger, "What Are Poets For?" p. 92.
11. Heidegger, "Building Dwelling Thinking," p. 148.
12. Ibid., p. 149.
13. Heidegger, "The Thing," p. 178.
14. Heidegger, "Building Dwelling Thinking," p. 149.
15. Heidegger, "The Thing," p. 178.
16. Vincent Vycinas, *Earth and Gods: An Introduction to the Philosophy of Martin Heidegger* (The Hague: Nijhoff, 1961).
17. Schurmann, *Heidegger on Being and Acting*, p. 171.
18. Heidegger, "Building Dwelling Thinking," p. 149.
19. Heidegger, "The Thing," p. 178.
20. Heidegger, "Building Dwelling Thinking," p. 150.
21. Ibid.
22. Heidegger, "The Thing," p. 178.
23. Martin Heidegger, "Poetically Man Dwells," in *Poetry, Language, Thought* (New York: Harper and Row, 1971), p. 220.
24. Jean Shinoda Bolen, *Goddesses in Everywoman: A New Psychology of Women* (New York: Harper Colophon Books, 1984), p. 46.
25. Jean Shinoda Bolen, *Gods in Everyman: A New Psychology of Men's Lives and Loves* (New York: Harper and Row, 1989), p. 162.
26. Bolen, *Goddesses in Everywoman*, p. 233.
27. Heidegger, "Building Dwelling Thinking," p. 150.
28. Heidegger, "Poetically Man Dwells," p. 220.
29. Heidegger, "Building Dwelling Thinking," p. 151.
30. Ibid., p. 152.
31. Ibid., p. 154.
32. Ibid., p. 160.
33. Ibid.
34. William Connolly, *Identity-Difference: Democratic Negotia-*

tions of Political Paradox (Ithaca, N.Y.: Cornell University Press, 1991), p. 14.

35. Perhaps such a way of living has broader consequences than is commonly thought. Thinking may communicate with other people in ways that are very mysterious. In *The Gods in Everyman* (p. 301), Jean Bolen relates the following discovery: "Off the shore in Japan, scientists had been studying monkey colonies on many separate islands for over thirty years. In order to keep track of the monkeys, they would drop sweet potatoes on the beach for them to eat. The monkeys would come out of the trees to get the sweet potatoes, and would be in plain sight to be observed. One day an 18-month-old female monkey named Imo started to wash her sweet potato in the sea before eating it. We can imagine that it tasted better without the grit and sand; maybe it even was slightly salty. Imo showed her playmates and her mother how to do it, and her friends showed their mothers, and gradually more and more monkeys began to wash their sweet potatoes instead of eating them grit and all. At first, only the adults who imitated their children learned, and gradually others did also. One day, the observers saw that all the monkeys on that particular island were washing their sweet potatoes.

"Although this was significant, what was even more fascinating to note was that when this shift happened, the behavior of the monkeys on all the other islands changed as well; they now all washed their sweet potatoes—despite the fact that monkey colonies on the different islands had no direct contact with each other."

Bolen believes that "the hundredth monkey" tipped the scales for culture. Just one more, and suddenly everything was different for all of them everywhere, in spite of being unconnected in ordinary ways. She uses this story as a myth to argue that in fact individual efforts can make big differences. Perhaps there is a group mind, a gestalt—or, dare we say, a god—that governs such things.

Bibliography

Ackermann, Robert. *Data, Instruments, and Theory: A Dialectical Approach to Understanding Science*. Princeton, N.J.: Princeton University Press, 1985.

Altizer, Thomas. *Total Presence: The Language of Jesus and the Language of Today*. New York: Seabury Press, 1980.

Arendt, Hannah. *Between Past and Future: Six Exercises in Political Thought*. Cleveland, N.Y.: Meridian Books, 1963.

———. *Eichmann in Jerusalem: A Report on the Banality of Evil*. New York: Penguin Books, 1971.

———. *The Human Condition*. Chicago: University of Chicago Press, 1958.

———. *The Origins of Totalitarianism*. New York: Harcourt Brace Jovanovich, 1973.

Aristotle. *Nichmachean Ethics*. Trans. Martin Ostwald. Indianapolis: Bobbs-Merrill, 1962.

———. *The Politics*. Trans. Ernest Barker. London: Oxford University Press, 1975.

Augustine. *City of God*. Ed. Vernon Bourke. New York: Image Book, 1958.

Avineri, Shlomo. *Hegel's Theory of the Modern State*. Cambridge: Cambridge University Press, 1972.

———. *The Social and Political Thought of Karl Marx*. Cambridge: Cambridge University Press, 1968.

Bahro, Rudolf. *Building the Green Movement*. Trans. Mary Tyler. Philadelphia: New Society Publishers, 1985.

Bamford, James. *The Puzzle Palace: A Report on NSA, America's Most Secret Agency*. Boston: Houghton Mifflin, 1982.

Barrett, William. *The Illusion of Technique*. London: William Kimber, 1979.

Barthes, Roland. *The Eiffel Tower*. Trans. Richard Howard. New York: Hill and Wang, 1979.

———. *Mythologies*. Trans. Annette Lavers. New York: Hill and Wang, 1972.

Bateson, Gregory. *Mind and Nature: A Necessary Unity*. New York: Bantam New Age Books, 1980.

Bell, Daniel. *The Cultural Contradictions of Capitalism*. New York: Basic Books, 1978.

Bellah, Robert N., Richard Madsen, William M. Sullivan, Ann Swidler, and Steven M. Tipton. *Habits of the Heart: Individualism and Commitment in American Life*. Berkeley: University of California Press, 1985.

Beniger, James R. *The Control Revolution: Technological and Economic Origins of the Information Society*. Cambridge: Harvard University Press, 1986.

Bennett, Jane. *Unthinking Faith and Enlightenment: Nature and State in a Post-Hegelian Era*. New York: New York University Press, 1987.

Berry, Thomas. *The Dream of the Earth*. San Francisco: Sierra Club, 1988.

Berry, Wendell. "The Body and the Machine." *Parabola: The Magazine of Myth and Tradition* 15, no. 3 (August 1990).

———. *The Country of Marriage*. San Diego: Harcourt Brace Jovanovich, 1973.

———. *Farming: A Hand Book*. New York: Harcourt Brace Jovanovich, 1970.

———. *The Gift of Good Land*. San Francisco: North Point Press, 1981.

———. *Recollected Essays, 1965–1980*. San Francisco: North Point Press, 1981.

———. *The Unsettling of America*. San Francisco: Sierra Club Books, 1977.

Bertalanffy, Ludwig von. *General System Theory*. New York: George Braziller, 1968.

Berlan, Jean-Pierre, and Richard Lewontin. "Technology, Research, and the Penetration of Capital: The Case of U.S. Agriculture." *Monthly Review* 38 (July–August 1986).

Best, Michael, and William Connolly. *The Politicized Economy*. Lexington, Mass.: D. C. Heath, 1982.

Bevis, William. *Ten Tough Trips: Montana Writers and the West.* Seattle: University of Washington Press, 1990.

Bimel, Walter. *Martin Heidegger: An Illustrated Study.* Trans. J. L. Mehta. New York: Harcourt Brace Jovanovich, 1976.

Blair, Bruce. *Strategic Command and Control: Redefining the Nuclear Threat.* Washington, D.C.: Brookings Institution, 1985.

Bloom, Harold, et al. *Deconstruction and Criticism.* New York: Continuum, 1979.

Bolen, Jean Shinoda. *Goddesses in Everywoman: A New Psychology of Women.* New York: Harper Colophon Books, 1984.

———. *Gods in Everyman: A New Psychology of Men's Lives and Loves.* New York: Harper and Row, 1989.

Bowles, Samuel, David Gordon, and Thomas Weisskopf. *Beyond the Wasteland.* Garden City, N.Y.: Anchor Press, 1983.

Bracken, Paul. *The Command and Control of Nuclear Forces.* New Haven, Conn.: Yale University Press, 1983.

Braudel, Fernand. *Capitalism and Material Life, 1400–1800.* New York: Harper and Row, 1967.

Braverman, Harry. *Labor and Monopoly Capital: The Degradation of Work in the Twentieth Century.* New York: Monthly Review Press, 1974.

Bright, James. *Automation and Management.* Boston: Maxwell, 1958.

———. "The Relationship of Increasing Automation and Skill Requirements." Appendix to *The Employment Impact of Technological Change*, vol. 2 of *Technology and the American Economy.* National Commission on Technology, Automation, and Economic Progress. Washington, D.C.: U.S. Government Printing Office, 1966.

Brown, Dee. *Bury My Heart at Wounded Knee.* New York: Pocket Books, 1981.

Brown, Lester R. *Building a Sustainable Society.* New York: W. W. Norton, 1981.

———. *State of the World, 1987.* New York: W. W. Norton, 1987.

———. *State of the World, 1989.* New York: W. W. Norton, 1989.

Campbell, Jeremy. *Grammatical Man: Information, Entropy, Language, and Life.* New York: Simon and Schuster, 1982.

Campbell, Joseph, with Bill Moyers. *The Power of Myth.* New York: Doubleday, 1988.

Camus, Albert. *The Rebel.* New York: Vintage Books, 1956.

Capra, Fritjof. *The Tao of Physics.* New York: Bantam Books, 1977.

Carter, Angela. *The Sadeian Woman and the Ideology of Pornography*. New York: Pantheon Books, 1978.
Carter, Vernon Gill, and Tom Dale. *Topsoil and Civilization*. Norman: University of Oklahoma Press, 1974.
Cavell, Stanley. *The World Viewed*. Cambridge: Harvard University Press, 1979.
Chodorow, Nancy. *The Reproduction of Mothering: Psychoanalysis and the Sociology of Gender*. Berkeley: University of California Press, 1978.
Chomsky, Noam, and Edward Herman. *The Washington Connection and Third World Fascism*. Boston: South End Press, 1979.
Commoner, Barry. *The Closing Circle*. New York: Bantam Books, 1977.
Connolly, William. *Appearance and Reality in Politics*. Cambridge: Cambridge University Press, 1981.
———. *Identity-Difference: Democratic Negotiations of Political Paradox*. Ithaca, N.Y.: Cornell University Press, 1991.
———. *Political Theory and Modernity*. New York: Basil Blackwell, 1988.
———. *Politics and Ambiguity*. Madison: University of Wisconsin Press, 1987.
Corlett, William. *Community without Unity*. Durham, N.C.: Duke University Press, 1989.
Cowan, Ruth Schwartz. *More Work for Mother*. New York: Basic Books, 1983.
Culler, Jonathan. *On Deconstruction: Theory and Criticism after Structuralism*. Ithaca, N.Y.: Cornell University Press, 1982.
———. *Structuralist Poetics*. Ithaca, N.Y.: Cornell University Press, 1975.
Dallmayr, Fred. *Margins of Political Discourse*. Albany: State University of New York Press, 1989.
———. "Ontology of Freedom: Heidegger and Political Philosophy." *Political Theory* 12, no. 2 (May 1984).
———. *Twilight of Subjectivity: Contributions to a Post-Individualist Theory of Politics*. Amherst: University of Massachusetts Press, 1981.
Daniels, Norman, ed. *Reading Rawls*. New York: Basic Books, 1980.
Davidson, Arnold I. "Symposium on Heidegger and Nazism." *Critical Inquiry* 15 (Winter 1989).
Deleuze, Gilles. *Nietzsche and Philosophy*. Trans. Hugh Tomlinson. New York: Columbia University Press, 1983.

De Man, Paul. *Allegories of Reading.* New Haven, Conn.: Yale University Press, 1979.

Demske, James M. *Being, Man, and Death: A Key to Heidegger.* Lexington: University Press of Kentucky, 1970.

Derrida, Jacques. "Heidegger, l'enfer des philosophes." *Le Nouvel Observateur,* November 6–12, 1987.

———. *Of Grammatology.* Trans. Gayatri C. Spivak. Baltimore: Johns Hopkins University Press, 1976.

———. *Of Spirit: Heidegger and the Question.* Trans. Geoffrey Bennington and Rachel Bowlby. Chicago: University of Chicago Press, 1989.

———. *Writing and Difference.* Trans. Alan Bass. Chicago: University of Chicago Press, 1978.

Descombes, Vincent. *Modern French Philosophy.* Cambridge: Cambridge University Press, 1979.

Dinnerstein, Dorothy. *The Mermaid and the Minotaur.* New York: Harper Colophon Books, 1977.

Dooling, D. M. "Focus." *Parabola* 16, no. 3 (August 1991).

Dossey, Larry. *Recovering the Soul: A Scientific and Spiritual Search.* New York: Bantam Books, 1989.

Douglas, Mary. *Purity and Danger.* London: Routledge, 1979.

Dreyfus, Hubert, and Paul Rabinow. *Michel Foucault: Beyond Structuralism and Hermeneutics.* Chicago: University of Chicago Press, 1982.

Dumm, Thomas L. *Democracy and Punishment: Disciplinary Origins of the United States.* Madison: University of Wisconsin Press, 1987.

Ehrlich, Paul, Carl Sagan, Donald Kennedy, and Walter Orr Roberts. *The Cold and the Dark: The World after Nuclear War.* New York: W. W. Norton, 1984.

Eiseley, Loren. *The Star Thrower.* New York: Harvest Books, 1978.

Eisler, Riane. *The Chalice and the Blade: Our History, Our Future.* San Francisco: Harper and Row, 1988.

Eliade, Mircea. *The Myth of the Eternal Return.* Princeton, N.J.: Princeton University Press, 1974.

Ellul, Jacques. *The Technological Society.* Trans. John Wilkinson. New York: Vintage Books, 1964.

Elshtain, Jean Bethke. *Public Man, Private Woman.* Princeton, N.J.: Princeton University Press, 1981.

———, ed. *The Family in Political Thought.* Amherst: University of Massachusetts Press, 1982.

Erikson, Erik. *Identity, Youth, and Crisis*. New York: W. W. Norton, 1968.

Farias, Victor. *Heidegger and Nazism*. Trans. Paul Burrell and Gabriel R. Ricci. Philadelphia: Temple University Press, 1989.

Fay, Brian. *Social Theory and Political Practice*. London: George Allen and Unwin, 1975.

Feyerabend, Paul. *Against Method*. London: Verso Editions, 1980.

———. *Science in a Free Society*. London: Verso Editions, 1978.

Ford, Daniel. *Meltdown: The Secret Papers of the Atomic Energy Commission*. New York: Simon and Schuster, Touchstone Books, 1986.

Foucault, Michel. *The Archaeology of Knowledge*. Trans. A. M. Sheridan Smith. New York: Harper and Row, 1972.

———. *Discipline and Punish*. Trans. Alan Sheridan. New York: Pantheon Books, 1977.

———. *The History of Sexuality*. Vol. I. Trans. Robert Hurley. New York: Random House, 1973.

———. *Madness and Civilization*. Trans. Richard Howard. New York: Vintage Books, 1973.

———. *The Order of Things*. Trans. Alan Sheridan. New York: Vintage Books, 1973.

———. *Power-Knowledge*. Trans. Colin Gordon, Leo Marshall, John Mepham, and Kate Soper. New York: Pantheon Books, 1977.

Freedman, Lawrence. *The Evolution of Nuclear Strategy*. New York: St. Martin's Press, 1983.

Freud, Anna. *The Ego and the Mechanisms of Defense*. Trans. Cecil Baines. New York: International Universities Press, 1966.

Freud, Sigmund. *Civilization and Its Discontents*. Trans. James Strachey. New York: W. W. Norton, 1961.

———. *The Future of an Illusion*. Trans. James Strachey. New York: W. W. Norton, 1961.

———. *General Psychological Theory*. Ed. Philip Rieff. New York: Collier Books, 1963.

———. *Group Psychology and the Analysis of the Ego*. Trans. James Strachey. New York: W. W. Norton, 1959.

———. *The Interpretation of Dreams*. Trans. James Strachey. New York: Avon Books, 1965.

———. *Introductory Lectures on Psychoanalysis*. Trans. James Strachey. New York: W. W. Norton, 1966.

———. *Moses and Monotheism*. Trans. Katherine Jones. New York: Vintage Books, 1967.

———. *New Introductory Lectures on Psychoanalysis*. Trans. James Strachey. New York: W. W. Norton, 1964.

———. *Studies in Parapsychology*. Ed. Philip Rieff. New York: Collier Books, 1971.

———. *Totem and Taboo*. Trans. James Strachey. New York: W. W. Norton, 1950.

Gadamer, Hans-Georg. "Back from Syracuse?" *Critical Inquiry* 15 (Winter 1989).

Galbraith, John Kenneth. *Economics and the Public Purpose*. New York: New American Library, 1973.

———. *The New Industrial State*. New York: New American Library, 1985.

Gibbons, Michael T. *Interpreting Politics*. New York: New York University Press, 1987.

Gillespie, Michael Allen. *Hegel, Heidegger, and the Ground of History*. Chicago: University of Chicago Press, 1984.

Girard, René. *The Scapegoat*. Trans. Yvonne Freccero. Baltimore: Johns Hopkins University Press, 1986.

Gleick, James. *Chaos: Making a New Science*. New York: Penguin Books, 1987.

Goodfield, June. *Playing God: Genetic Engineering and the Manipulation of Life*. New York: Random House, 1977.

Gorz, André. *Ecology as Politics*. Trans. Patsy Vigderman and Jonathan Cloud. Boston: South End Press, 1980.

Goudsblom, Johan. *Nihilism and Culture*. Totowa, N.J.: Rowman and Littlefield, 1980.

Gould, Stephen Jay. *Ever Since Darwin*. New York: W. W. Norton, 1977.

———. *The Flamingo's Smile*. New York: W. W. Norton, 1985.

———. *Hen's Teeth and Horse's Toes*. New York: W. W. Norton, 1983.

———. *The Mismeasure of Man*. New York: W. W. Norton, 1981.

———. *The Panda's Thumb*. New York: W. W. Norton, 1982.

Greenberg, Daniel. *The Politics of Pure Science*. New York: New American Library, 1967.

Griffin, Susan. *Women and Nature*. New York: Harper Colophon Books, 1978.

Gutierrez, Gustavo. *A Theology of Liberation*. Maryknoll, N.Y.: Orbis Books, 1973.

Gyorgy, Anna, and friends. *No Nukes: Everyone's Guide to Nuclear Power*. Boston: South End Press, 1979.

Habermas, Jürgen. *Legitimation Crisis.* Boston: Beacon Press, 1973.

Hamilton, Alexander, James Madison, and John Jay. *The Federalist Papers.* New York: New American Library, 1961.

Hampshire, Stuart. *Freedom of the Individual.* Princeton, N.J.: Princeton University Press, 1977.

Hanson, Dirk. *The New Alchemists.* New York: Avon Books, 1982.

Haraway, Donna J. *Semians, Cyborgs, and Women: The Reinvention of Nature.* New York: Routledge, 1991.

Hegel, G. W. F. *Phenomenology of the Spirit.* Trans. A. V. Miller. Oxford: Oxford University Press, 1979.

———. *Philosophy of Right.* Trans. T. M. Knox. London: Oxford University Press, 1979.

———. *Reason in History.* Trans. Robert S. Hartman. Indianapolis: Bobbs-Merrill, 1953.

Heidegger, Martin. *Being and Time.* Trans. John Macquarrie and Edward Robinson. New York: Harper and Row, 1962.

———. *Discourse on Thinking.* Trans. John M. Anderson and E. Hans Freund. New York: Harper and Row, 1966.

———. *Identity and Difference.* Trans. Joan Stambaugh. New York: Harper and Row, 1982.

———. *An Introduction to Metaphysics.* Trans. Ralph Manhiem. New Haven, Conn.: Yale University Press, 1976.

———. *Martin Heidegger: Basic Writings.* Ed. David Krell. New York: Harper and Row, 1977.

———. *Nietzsche.* Vols. 1, 3, and 4. Ed. David Krell. San Francisco: Harper and Row, 1979.

———. *On the Way to Language.* Trans. Peter D. Hertz. New York: Harper and Row, 1982.

———. *On Time and Being.* Trans. Joan Stambaugh. New York: Harper and Row, 1972.

———. *Poetry, Language, Thought.* Trans. Albert Hofstadter. New York: Harper and Row, 1971.

———. *The Question concerning Technology.* Trans. William Lovitt. New York: Harper and Row, 1977.

———. "The Self-Assertion of the German University: Address Delivered on the Solemn Assumption of the Rectorate of the University of Freiburg" and "The Rectorate, 1933/34: Facts and Thoughts." Trans. Karsten Harries. *Review of Metaphysics* 38 (March 1985).

———. *What Is a Thing*. Trans. W. B. Barton, Jr., and Vera Deutsch. South Bend, Ind.: Gateway Editions, 1976.

———. *What Is Called Thinking?* Trans. J. Glenn Gray. New York: Harper and Row, 1968.

Heisenberg, Werner. *Across the Frontiers*. New York: Harper and Row, 1974.

Heller, Peter. *Dialectics and Nihilism: Essays on Lessing, Nietzsche, Mann, and Kafka*. Amherst: University of Massachusetts Press, 1966.

Herman, Edward. *The Real Terror Network*. Boston: South End Press, 1982.

Hirsch, Fred. *Social Limits to Growth*. Cambridge: Harvard University Press, 1978.

Hirschman, Albert O. *The Passions and the Interests: Political Arguments for Capitalism before Its Triumph*. Princeton, N.J.: Princeton University Press, 1977.

Hobbes, Thomas. *Leviathan*. Parts 1 and 2. Indianapolis: Bobbs-Merrill, 1958.

Hofstadter, Douglas R. *Gödel, Escher, Bach: An Eternal Golden Braid*. New York: Random House, 1980.

Horwitz, Robert, ed. *The Moral Foundations of the American Republic*. Charlottesville: University Press of Virginia, 1979.

Hyams, Edward. *Soil and Civilization*. New York: Harper and Row, 1976.

Illich, Ivan. *Toward a History of Needs*. New York: Bantam New Age, 1977.

Jahoda, Marie. *Freud and the Dilemmas of Psychology*. New York: Basic Books, 1977.

Jaspers, Karl. *Nietzsche*. Trans. Charles F. Walraff and Frederick J. Schmitz. South Bend, Ind.: Gegnery/Gateway, 1965.

Johnson, Barbara. *A World of Difference*. Baltimore: Johns Hopkins University Press, 1989.

Journal of the New Alchemists. All issues. Brattleboro, Vt.: Stephen Greene Press.

Jung, Carl G. *Collected Works*. Executive ed., William McGuire. Princeton, N.J.: Princeton University Press, 1990.

Kaku, Michio, and Daniel Axelrod. *To Win a Nuclear War*. Boston: South End Press, 1987.

Kant, Immanuel. *Foundations of the Metaphysics of Morals*. Trans. Lewis White Beck. Indianapolis: Bobbs-Merrill, 1959.

———. *Prolegomena to Any Future Metaphysics.* Trans. Lewis White Beck. Indianapolis: Bobbs-Merrill, 1950.

Kariel, Henry. *The Desperate Politics of Postmodernism.* Amherst: University of Massachusetts Press, 1989.

Kaufmann, Walter. *Nietzsche: Philosopher, Psychologist, Anarchist.* New York: Vintage Books, 1968.

Keenan, Tom. "The 'Paradox' of Knowledge and Power: Reading Foucault on a Bias." *Political Theory* 15, no. 1 (February 1987).

Keisling, Bill. *Solar Water Heating Systems.* Emmaus, Pa.: Rodale Press, 1983.

Kierkegaard, Søren. *Fear and Trembling and the Sickness unto Death.* Trans. Walter Lowrie. Princeton, N.J.: Princeton University Press, 1954.

Kissinger, Henry. *Nuclear Weapons and Foreign Policy.* New York: W. W. Norton, 1958.

Kockelmans, Joseph. *Heidegger and Science.* Washington, D.C.: University Press of America, 1985.

———. *On Heidegger and Language.* Evanston, Ill.: Northwestern University Press, 1972.

———. *On the Truth of Being.* Bloomington: Indiana University Press, 1984.

Kojeve, Alexandre. *Introduction to the Reading of Hegel.* Ed. Allan Bloom. Ithaca, N.Y.: Cornell University Press, 1969.

Komarov, Boris. *The Destruction of Nature in the Soviet Union.* Trans. Michel Vale and Joe Hollander. White Plains, N.Y.: M. E. Sharpe, 1980.

Kovel, Joel. *Against the State of Nuclear Terror.* Boston: South End Press, 1983.

Kovesi, Julius. *Moral Notions.* New York: Routledge, 1967.

Kramer, Heinrich, and James Sprenger. *The Malleus Maleficarum.* Trans. Montague Sommers. New York: Dover Publications, 1971.

Krell, David F. *Postponements: Women, Sensuality, and Death in Nietzsche.* Bloomington: Indiana University Press, 1986.

Kuhn, Thomas. *The Essential Tension.* Chicago: University of Chicago Press, 1962.

———. *The Structure of Scientific Revolutions.* Chicago: University of Chicago Press, 1962.

Kundera, Milan. *The Unbearable Lightness of Being.* New York: Harper and Row, 1984.

Kurzweil, Edith. *The Age of Structuralism.* New York: Columbia University Press, 1982.

Lacoue-Labarthe, Philippe. *Heidegger, Art, and Politics.* Trans. Chris Turner. Cambridge: Basil Blackwell, 1990.

Lacoue-Labarthe, Philippe, and Jean-Luc Nancy, eds. *Les fins de l'homme: A partir du travail de Jacques Derrida.* Paris: Editions Galilée, 1980.

Lang, R. D. *The Politics of Experience.* New York: Ballantine Books, 1968.

Lauer, Quentin. *A Reading of Hegel's Phenomenology of Spirit.* New York: Fordham University Press, 1982.

Leiss, William. *The Domination of Nature.* New York: Beacon Press, 1974.

Leitch, Vincent. *Deconstructive Criticism: An Advanced Introduction.* New York: Columbia University Press, 1983.

Lenin, V. I. *Collected Works.* Moscow: International Press, 1965.

LeShan, Lawrence. *The Medium, the Mystic, and the Physicist.* New York: Viking Press, 1974.

Levin, David Michael. *The Opening of Vision: Nihilism and the Postmodern Situation.* New York: Routledge, 1988.

Lovelock, J. E. *Gaia: A New Look at Life on Earth.* Oxford: Oxford University Press, 1979.

Lovins, Amory B. *Soft Energy Paths: Toward a Durable Peace.* New York: Harper and Row, 1977.

Lovins, Amory B., and Hunter Lovins. *Brittle Power: Energy Strategy for National Security.* Andover, Mass.: Brick House, 1982.

Luke, Timothy W. "What's Wrong with Deterrence? A Semiotic Interpretation of National Security Policy." In *International-Intertextual Relations: Postmodern Readings of World Politics*, ed. James Der Derian and Michael Shapiro. Lexington, Mass.: Lexington Books, 1989.

McCloskey, Donald. *The Rhetoric of Economics.* Madison: University of Wisconsin Press, 1985.

MacIntyre, Alasdair. *After Virtue.* Notre Dame, Ind.: University of Notre Dame Press, 1981.

McLuhan, T. C., comp. *Touch the Earth.* New York: Simon and Schuster, Touchstone Books, 1971.

McMillan, Carol. *Women, Reason, and Nature.* Princeton, N.J.: Princeton University Press, 1982.

McRobie, George. *Small Is Possible.* New York: Harper and Row, 1981.

McWilliams, Carey. *Factories in the Field.* Santa Barbara, Calif.: Peregrine Publishers, 1971.

Martindale, David. *Earth Shelters.* New York: E. P. Dutton, 1981.
Marx, Karl. *Capital.* Vol. 1. Ed. Frederick Engels. New York: International Publishers, 1977.
———. *The Marx-Engels Reader.* Ed. Robert Tucker. New York: W. W. Norton, 1978.
Marx, Werner. *Heidegger and the Tradition.* Evanston, Ill.: Northwestern University Press, 1971.
Mayer, Milton. *They Thought They Were Free.* Chicago: University of Chicago Press, 1955.
Mazria, Edward. *The Passive Solar Energy Book.* Emmaus, Pa.: Rodale Press, 1979.
Megill, Allan. *Prophets of Extremity: Nietzsche, Heidegger, Foucault, Derrida.* Berkeley: University of California Press, 1985.
Metzner, Ralph. "Getting to Know One's Inner Enemy." In *Human Survival and Consciousness Evolution*, ed. Stanislov Grof. Albany: State University of New York Press, 1988.
Morgan, Robin. *The Anatomy of Freedom: Feminism, Physics, and Global Politics.* New York: Anchor Press, 1982.
Mosse, George. *Toward the Final Solution: A History of European Racism.* New York: Harper Colophon Books, 1970.
Mumford, Lewis. *The Condition of Man.* New York: Harcourt Brace, 1944.
———. *The Pentagon of Power: The Myth of the Machine.* New York: Harcourt Brace Jovanovich, 1970.
———. *Technics and Civilization.* New York: Harcourt Brace Jovanovich, 1962.
Murray, Michael, ed. *Heidegger and Modern Philosophy.* New Haven, Conn.: Yale University Press, 1978.
Neihardt, John. *Black Elk Speaks.* New York: Washington Square Press, 1959.
Niebuhr, Reinhold. *Beyond Tragedy.* New York: Charles Scribner's Sons, 1937.
Nietzsche, Friedrich. *The Birth of Tragedy and the Genealogy of Morals.* Trans. Francis Golffing. Garden City, N.Y.: Doubleday, 1956.
———. *On the Genealogy of Morals and Ecce Homo.* Ed. Walter Kaufmann. New York: Random House, 1969.
———. *The Portable Nietzsche.* Ed. Walter Kaufmann. New York: Viking Books, 1968.
———. *Thus Spoke Zarathustra.* Trans. R. J. Hollingdale. New York: Penguin Books, 1969.

———. *Twilight of the Idols and the Anti-Christ.* Trans. R. J. Hollingdale. Baltimore: Penguin Books, 1968.

———. *The Will to Power.* Ed. Walter Kaufmann. New York: Vintage Books, 1974.

Nimmo, Dan. *The Political Persuaders: The Techniques of Modern Election Campaigns.* Englewood Cliffs, N.J.: Prentice-Hall, 1970.

Noble, David. *America by Design: Science, Technology, and the Rise of Corporate Capitalism.* Oxford: Oxford University Press, 1977.

Norman, Colin. *The God That Limps: Science and Technology in the Eighties.* New York: W. W. Norton, 1981.

Norton, Anne. *Reflections on Political Identity.* Baltimore: Johns Hopkins University Press, 1988.

O'Connor, James. *The Fiscal Crisis of the State.* New York: St. Martin's Press, 1973.

Oelschlaeger, Max. *The Idea of Wilderness.* New Haven, Conn.: Yale University Press, 1991.

Ortner, Sherry. *High Religion: A Cultural and Political History of Sherpa Buddhism.* Princeton, N.J.: Princeton University Press, 1989.

Oski, Frank. "Formula for Profits: Heating Up the Bottle Battle." *Nation* 249, no. 19 (December 4, 1989).

Pagels, Elaine. *Adam, Eve, and the Serpent.* New York: Random House, 1988.

Pearson, Jeff, and Jessica Pearson. *No Time But Place: A Prairie Pastoral.* New York: McGraw-Hill, 1980.

Peck, M. Scott. *People of the Lie.* New York: Simon and Schuster, Touchstone Books, 1983.

Peterson, Aage. *Quantum Physics and the Philosophical Tradition.* Cambridge: MIT Press, 1968.

Pirsig, Robert. *Zen and the Art of Motorcycle Maintenance.* New York: Bantam Books, 1975.

Plato. *The Republic.* Trans. Francis Macdonald. Oxford: Oxford University Press, 1941.

Platon, Nicolas. *Crete.* Geneva: Nagel Publishers, 1966.

Raschke, Carl A. *The Bursting of New Wine Skins: Reflections on Religion and Culture at the End of Affluence.* Pittsburgh: Pickwick Press, 1971.

Ravetz, Jerome. *Scientific Knowledge and Its Social Problems.* New York: Oxford University Press, 1971.

Rawls, John. *A Theory of Justice*. Cambridge: Harvard University Press, 1971.

Ricoeur, Paul. *Freud and Philosophy: An Essay on Interpretation*. New Haven, Conn.: Yale University Press, 1970.

Rochester, Anna. *Why Farmers Are Poor*. New York: International Publishers, 1940.

Rogin, Michael. *Ronald Reagan the Movie, and Other Episodes in Political Demonology*. Berkeley: University of California Press, 1987.

Rorty, Richard. *Philosophy and the Mirror of Nature*. Princeton, N.J.: Princeton University Press, 1982.

Rosen, Stanley. *Nihilism: A Philosophical Essay*. New Haven, Conn.: Yale University Press, 1969.

Rousseau, Jean-Jacques. *First and Second Discourses*. Ed. Roger Masters. New York: St. Martin's Press, 1964.

———. *On the Social Contract: With Geneva Manuscript and Political Economy*. Ed. Roger Masters, trans. Judith Masters. New York: St. Martin's Press, 1978.

Roy, Robert L. *Underground Houses*. New York: Sterling Publishing, 1982.

Rubenstein, Diane. *Whats Left? The Ecole Normale Supérieure and the Right*. Madison: University of Wisconsin Press, 1990.

Rubin, Lillian Breslow. *Worlds of Pain: Life in the Working-Class Family*. New York: Basic Books, 1977.

Sade, The Marquis de. *Juliette*. Trans. Austryne Wainhouse. New York: Grove Press, 1968.

———. *The Marquis de Sade*. Trans. Richard Seaver and Austryn Wainhouse. New York: Grove Press, 1965.

———. *The 120 Days of Sodom*. Trans. Richard Seaver and Austryn Wainhouse. New York: Grove Press, 1966.

Salvadori, Mario. *Why Buildings Stand Up: The Strength of Architecture*. New York: W. W. Norton, 1980.

Sampson, Neil R. *Farmland or Wasteland: A Time to Choose*. Emmaus, Pa.: Rodale Press, 1981.

Schell, Jonathan. *The Fate of the Earth*. New York: Avon, 1982.

———. *The Time of Illusion*. New York: Random House, 1976.

Schirmacher, Wolfgang. *Technik und Gelassenheit*. Freiberg: Karl Alber, 1984.

Schrecker, Ellen. *No Ivory Tower: McCarthyism and the Universities*. New York: Oxford University Press, 1986.

Schultze, Charles L. *The Public Use of Private Interest.* Washington, D.C.: Brookings Institution, 1977.
Schumacher, E. F. *Good Work.* New York: Harper and Row, 1977.
———. *Small Is Beautiful: Economics As If People Mattered.* New York: Harper and Row, 1977.
Schurmann, Reiner. *Heidegger on Being and Acting: From Principles to Anarchy.* Trans. Christine-Marie Gros. Bloomington: Indiana University Press, 1987.
Seigel, Jerrold. *Marx's Fate: The Shape of a Life.* Princeton, N.J.: Princeton University Press, 1978.
Sennett, Richard, and Jonathan Cobb. *The Hidden Injuries of Class.* New York: Vintage Books, 1973.
Serres, Michel. *The Parasite.* Trans. Lawrence R. Schehr. Baltimore: Johns Hopkins University Press, 1982.
Shapiro, Michael. *Language and Political Understanding: The Politics of Discursive Practices.* New Haven, Conn.: Yale University Press, 1981.
Sheehan, Neil. *A Bright Shining Lie: John Paul Vann and America in Vietnam.* New York: Random House, 1988.
Sheldrake, Rupert. *The Presence of the Past: Morphic Resonance and the Habits of Nature.* New York: Vintage Press, 1989.
Shelley, Mary. *Frankenstein; or, the Modern Prometheus.* New York: New American Library, 1965.
Snyder, Gary. *Turtle Island.* New York: New Directions, 1974.
Spriegel, William R., and Clark E. Myers, eds. *The Writings of the Gilbreths.* Homewood, Ill.: R. D. Irwin, 1953.
Steiner, George. *Martin Heidegger.* New York: Penguin Books, 1962.
Stone, Merlin. *When God Was a Woman.* New York: Harcourt Brace Jovanovich, 1976.
Sweezy, Paul. *The Theory of Capitalist Development.* New York: Modern Reader, 1970.
Taylor, Charles. *Hegel.* Cambridge: Cambridge University Press, 1973.
Taylor, Frederick W. *The Principles of Scientific Management.* New York: W. W. Norton, 1967.
Taylor, Mark C. *Deconstructing Theology.* Chico, Calif.: Scholars Press, 1982.
———. *Erring: A Postmodern A-Theology.* Chicago: University of Chicago Press, 1984.
Thoreau, Henry David. *Walden and Other Writings.* New York: Bantam Books, 1962.

Toffler, Alvin. *The Eco-Spasm Report.* New York: Bantam Books, 1975.
Underground Space Center, University of Minnesota. *Earth Sheltered Housing Design: Guidelines, Examples, and References.* New York: Van Nostrand Reinhold, 1978.
U.S. Bureau of the Census. *Statistical Abstract of the United States, 1982–83.* Washington, D.C.: U.S. Government Printing Office, 1982.
U.S. Department of Agriculture. *The 1984 Fact Book of U.S. Agriculture.* Washington, D.C.: U.S. Government Printing Office, 1984.
Vacca, Roberto. *The Coming Dark Age.* New York: Anchor Books, 1974.
Vahanian, Gabriel. *The Death of God: The Culture of Our Post-Christian Era.* New York: George Braziller, 1961.
———. *Wait without Idols.* New York: George Braziller, 1964.
Vycinas, Vincent. *Earth and Gods: An Introduction to the Philosophy of Martin Heidegger.* The Hague: Nijhoff, 1961.
Wade, Alex, and Neal Ewenstein. *Thirty Energy-Efficient Houses You Can Build.* Emmaus, Pa.: Rodale Press, 1977.
Watts, Alan. *Nature, Man, and Woman.* New York: Vintage Books, 1970.
Weatherford, Jack. *Indian Givers: How the Indians of the Americas Transformed the World.* New York: Fawcett Columbine, 1988.
Whitmont, Edward. *Return of the Goddess.* New York: Crossroad Publishing, 1982.
Wilber, Ken, ed. *The Holographic Paradigm and Other Paradoxes.* Boulder, Colo.: Shambhala, 1987.
Wills, Gary. *Explaining America: The Federalist.* New York: Penguin Books, 1981.
Wollheim, Richard, and James Hopkins, eds. *Philosophical Essays on Freud.* Cambridge: Cambridge University Press, 1982.
Woodward, Bob. *The Commanders.* New York: Simon and Schuster, 1991.
Wright, Erik Olin. *Class, Crisis and the State.* London: NLB, 1978.
Zimmerman, Michael E. "The Thorn in Heidegger's Side: The Question of National Socialism." *Philosophical Forum* 20, no. 4 (Summer 1989).

Index

About the Author

Wade Sikorski is an independent scholar who lives in Willard, Montana. He received his bachelor's degree from Montana State University and his doctorate from the University of Massachusetts (Amherst).